T0190096

Textile Science and Clothing Technology

Series editor

Subramanian Senthilkannan Muthu, Kowloon, Hong Kong

More information about this series at http://www.springer.com/series/13111

Subramanian Senthilkannan Muthu
Editor

Sustainable Innovations in Textile Fibres

Editor
Subramanian Senthilkannan Muthu
Kowloon
Hong Kong

ISSN 2197-9863 ISSN 2197-9871 (electronic)
Textile Science and Clothing Technology
ISBN 978-981-13-4189-2 ISBN 978-981-10-8578-9 (eBook)
https://doi.org/10.1007/978-981-10-8578-9

Printed on acid-free paper

This Springer imprint is published by the registered company Springer Nature Singapore Pte Ltd.
part of Springer Nature
The registered company address is: 152 Beach Road, #21-01/04 Gateway East, Singapore 189721, Singapore

This book is dedicated to:
The lotus feet of my beloved
Lord Pazhaniandavar
My beloved late Father
My beloved Mother
My beloved Wife Karpagam and
Daughters—Anu and Karthika
My beloved Brother
Last but not least
To everyone working in the global apparel
sector to make it SUSTAINABLE

Contents

Sustainable Textile Fibers

S. Grace Annapoorani

Abstract In the present period of ecological awareness an ever increasing number of materials are rising around the world, productive use of plant species and using the finest particles and filaments acquired from different eco cordial materials. Material is the real piece of the fundamental human needs. Ecological effects happen at each phase of the life cycle of an item. Common strands are fundamentally lengthened substances delivered by plants and creatures that can be spun into fibers, yarns and ropes. Like horticulture, materials have been a principal part of human life since the beginning of development. In India, a developing lack of normal fiber makers drives the scientists to grow new natural cordial material and its items. Natural fibres are at the heart of an eco-fashion movement that seeks to create garments that are sustainable at every stage of their life cycle, from production to disposal. Natural fibres have intrinsic properties such as mechanical strength, low weight and healthier to the wearer that has made them particularly attractive. Natural fibres have inherent properties, for example, mechanical quality, low weight and more beneficial to the wearer that has made them especially alluring. Logically, eco-materials are being utilized for mechanical purposes and additionally in segments of composite materials, in restorative inserts. Common and natural apparel is charged more by the retailers, since the wellspring of the fiber are free from herbicides, pesticides, or innately adjusted for a situation and these filaments procedure are not polished on an extensive scale. Due to all these they turn into somewhat costly. Be that as it may, while wearing, one can feel the fascinating extravagance making its cost irrelevant. The textures made out of eco filaments can be worn by any one as they don't have any b"ing chemicals in them. Hence the usage of eco fibers and organic are the best solution to keep our earth clean and to minimize the global warming. At present, the utilization and disposal of the textile will be more of environmental sustainable to minimize harm to people and the environment.

Keywords Natural fibers · Organic ecofibers · Ecofriendly · Sustainable ecofriendly fibers

---

S. Grace Annapoorani (✉)
Department of Textiles and Apparel Design, Bharathiar University, Coimbatore, Tamil Nadu, India
e-mail: gracetad@buc.edu.in

S. S. Muthu (ed.), *Sustainable Innovations in Textile Fibres*, Textile Science and Clothing Technology, https://doi.org/10.1007/978-981-10-8578-9_1

1 Introduction

There are three essential needs that a man has food, clothing and shelter. The worldwide material and textile industry will undoubtedly be colossal, as it satisfies the second fundamental prerequisite of man. Everyone needs to hit an impression with various and popular garments. Yet, the tragic actuality is that the human eagerness to look engaging and wear glitzy garments has wound up making hurt nature. The textile industry is a standout amongst the most contaminations discharging industry of the world. Overviews demonstrate that about 5% of all landfill space is devoured by material waste. Also, 20% of all fresh water contamination is made by material treatment and dyeing.

Poisons discharged by the worldwide textile industry are constantly doing inconceivable damage to nature. It dirties land and makes them useless and infertile over the long run. Overviews demonstrate that cotton devours the most elevated measure of unsafe pesticides and composts. The textile industry utilizes a large number of gallons of water each day. The issue does not rest in the high use; however the waste is not dealt with to remove contaminations from it before it is arranged to water bodies.

The fluid effluents discharged by the textile industry are the most exasperating region of concern. This is on the grounds that the toxic material discharged through fluid waste is immense in quantity. It comprises of chemicals, for example, formaldehyde (HCHO), chlorine, and substantial metals. Additionally it is arranged into water bodies that range far away zones and is devoured by an expansive number of individuals for drinking or for day by day activities.

Air contamination caused by the textile industry is likewise a noteworthy reason for concern. Boilers, thermo pack, and diesel generators create poisons that are discharged into the air. The contaminations created incorporate Suspended Particulate Matter (SPM), sulfur di-oxide gas, oxide of nitrogen gas, and so on. The nearby zones with human population get influenced unfavorably inferable from the arrival of harmful gas into the environment.

It has turned out to be completely important to decrease the pollutants discharged by the textile industry. Polluting of the air, water, and land by textile industry and its raw material assembling units has turned into a genuine risk to the earth. It has endangered the life of individuals and different species on Earth. A dangerous atmospheric devotion is an immediate aftereffect of the toxins discharged by such industries. It likewise causes unsafe infections and health problems in individuals getting presented to the contaminations over the long run.

The utilization of natural raw material can help in battling the discharge of contaminations by the textile units. Organic cotton is particularly valuable as the creation of cotton requests the most extreme measure of pesticides and manures. In addition, the waste created from textile assembling plants should to be handled in a way that it is free from toxic chemicals previously it is disposed. Condition well-disposed strategies for development and produce should to be depended on. The word "eco" is short word used for ecology. Ecology is the investigation of the cooperation's amongst living beings and their condition. Thus ecology friendly is a term to refer to merchandise and serves considered to perpetrate negligible or no damage on the earth.

2 Textile Industry Overview

India is the second largest producer of textile materials in the world. Abundant accessibility of raw materials, for example, cotton, fleece, silk and jute have made the nation a sourcing center.

The textile industry has made a noteworthy commitment to the national economy regarding immediate and backhanded work age and net remote trade profit. It gives guide work to more than 45 million individuals. The textiles division is the second biggest supplier of work after farming. Consequently, development and all round advancement of this industry has an immediate bearing on the change of the India's economy.

The Indian textile industry is set for solid development, floated by solid residential utilization and also trade request. The key quality of the textile industry streams from its solid generation base of extensive variety of fibers/yarns from regular fibers like cotton, jute, silk and fleece to engineered/man-made fibers to like polyester, thick, nylon and acrylic.

2.1 Factors Favoring Growth of the Indian Textile Industry

1. **Raw material base**: India has high independence for raw material especially natural fibers. India's cotton raw material is the third biggest on the world. Indian textile Industry delivers and handles a wide range of fibers.
2. **Labor**: Cheap work and solid entrepreneurial aptitudes have dependably been the foundation of the Indian textile Industry.
3. **Flexibility**: The little size of assembling which is prevalent in the attire industry considers more prominent adaptability to benefit littler and specific requests.
4. **Rich heritage**: The social decent variety and rich legacy of the nation offers great motivation base for producer.
5. **Domestic market**: Natural request drivers including rising pay levels, expanding urbanization and development of the buying populace drive residential request.

India has been notable for her textile merchandise since extremely old circumstances. The conventional textile industry of India was practically rotted during the frontier administration. Be that as it may, the advanced textile industry took birth in India in the mid nineteenth century when the primary textile factory in the nation was set up at fortress gloater close Calcutta in 1818. The cotton textile industry, be that as it may, made its genuine start in Bombay, in 1850s. The principal cotton textile plant of Bombay was set up in 1854 by a Parsi cotton vendor at that point occupied with textile mill and inward exchange. To be sure, by far most of the early plants were the craftsmanship of Parsi vendors occupied with yarn and fabric exchange at home and Chinese and African markets.

The first cotton process in Ahmedabad, which was in the long run to develop as an opponent Center to Bombay, was built up in 1861. The spread of the textile industry to Ahmedabad was to a great extent because of the Gujarati exchanging class.

The cotton textile industry gained quick ground in the second 50% of the nineteenth century and before the centuries over there were 178 cotton textile mills; yet during the year 1900 the cotton textile industry was in awful state because of the immense starvation and various factories of Bombay and Ahmedabad were to be shut down for long stretches.

3 Textile Production and the Environment

The homesteads that develop raw materials used to make fabrics, including crops like flax, hemp and cotton, require a great deal of water. Truth be told, cotton is a particularly thristy plant. Likewise, to secure these important harvests, cultivates additionally utilize loads of pesticides and herbicides that at that point wind up in the environment. Once more, cotton is a major guilty party, being a standout amongst the most pesticide-escalated trims on the world.

Different types of materials additionally go through a considerable measure of characteristic assets. Assembling rayon, a counterfeit fabric produced using wood pulp, has brought about the loss of numerous old-development backwoods. What's more, during the procedure that changes it into fabric, the pulp is treated with risky chemicals that in the long run discover their way into the environment.

Presently consider manufactured or man-made fabrics to like nylon and polyester. These textiles are produced using petrochemicals and petroleum derivatives, and making them requires loads of water and vitality. Nylon fabricating additionally makes nursery gasses that mischief the air we relax. What's more, engineered fabrics are not biodegradable, which implies something made of nylon can take a very long time to break down.

Making textiles likewise includes exercises like dying, coloring, and washing that utilization lots of water. Such procedures deliver salts, surfactants, which enable colors to enter fabrics, and other surface-dynamic operators like cleanser that don't break down, so they wind up in our water. Coloring and printing likewise now and again include risky chemicals and substances like lead, mercury, and arsenic.

The textile industry has been denounced as being one of the world's most noticeably awful guilty parties as far as contamination. Generally created fabrics contain residuals of chemicals utilized during their Manufacture—chemicals that vanish into the air we inhale or are consumed through our skin. A portion of the chemicals are cancer-causing or may make hurt youngsters even before birth while others may trigger unfavorably susceptible responses in a few people. Cotton is the second-most harming horticultural yield on the world; 25% of all Pesticides utilized comprehensively are put on cotton crops. Majority of the cotton is flooded and the mix of substance application (through pesticides and manures) with water system is an immediate channel for harmful chemicals to flow in groundwater around the world.

There are two fundamental types of fibers used to make materials—Natural and Synthetic. Natural Fibers as a result of their starting point are bio degradable and innocuous to the earth, on the condition they are handled without the utilization of chemicals. Organic Farming is an exorbitant undertaking as it includes more care and no utilization of fake intends to develop the products, the deliver too is low when contrasted with industry cultivating, and thus naturally developed fibers are costly.

The Textile Industry includes parcel of procedures ideal from the development of the fibers to the last phase of fabric. The Spinning, Weaving and the Processing industry produces parcel of unsafe squanders hampering our environment.

3.1 Environment Impacts

Effect on Environment any modern movement causes contamination in one form or the other as is the textile industry. The effect of textile production on the natural viewpoints, for example, air, water, land and human body, and the social perspectives, for example, kid work and poor unhygienic working conditions must be considered. As of late, another measurement is presented for the earth agreeableness of the completed item. This incorporates the prohibition on certain azo colors, which are known or suspected to be cancer-causing, and the nearness of unsafe chemicals, (for example, formaldehyde) and certain metals. Some of these angles are quickly talked about. In spite of the fact that there are natural dangers during the whole creation chain (Fig. 1), the material wet handling has genuine natural problems. Substantial number of chemicals in huge amounts is utilized as a part of material wet preparing to fulfill buyer's requests as respects feel, handle, bestowing attractive properties, and so forth. Some of these chemicals, for example, colors and completing specialists, stay joined to the materials, while a generous extent of these chemicals stay in the handled water, causing air and water contamination. Air contamination is likewise caused during drying and polymerization grouping of completing operations. A considerable lot of the colors and completing operators staying on the completed fabric have been found to posture wellbeing dangers. Reports about instances of wellbeing harm with slogans "harm in closet" have extraordinarily activated the general assessment in created nations. The common production analysis for cotton wet processing is shown in Fig. 1.

India is the principal nation that has coordinated the security and change of the environment in its constitution. The different ecological enactment and controls managing insurance and change of the environment are quickly recorded underneath. There are no particular natural laws for material industry segment alone. In any case, there are industry particular guidelines, which the textile industry is required to consent to while setting up or working a modern unit. The administrative specialists are

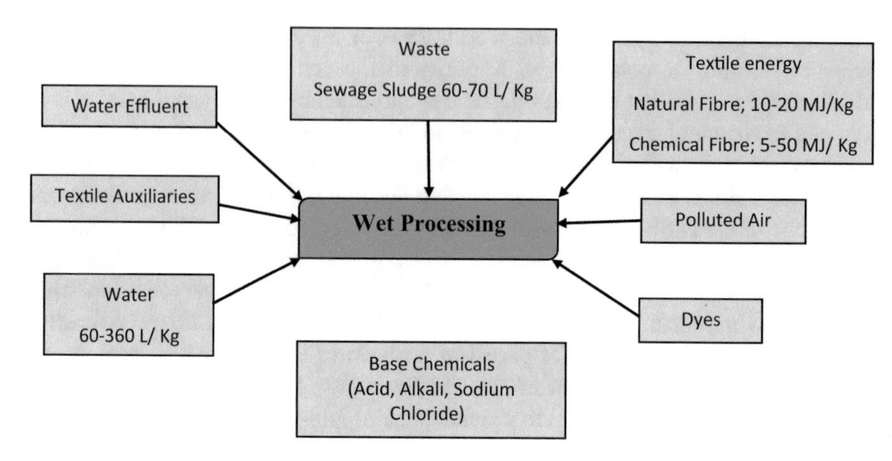

Fig. 1 Wet processing process. *Source* http://fashionarun.page.tl/ECO-FRIENDLY-TEXTILES.html

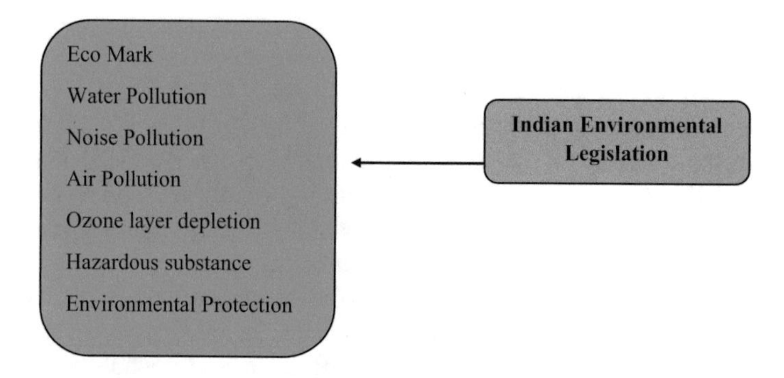

Fig. 2 Types of pollutions. *Source* www.environmentalpollution.in

Ministry of Environment and Forests (MoEF) and Central Pollution Control Board (CPCB) at central level and State Pollution Control Board (SPCB) at state level. Requirement is finished by SPCBs. Types of pollutions is shown in Fig. 2.

3.2 Environment Protection Act

The accompanying are the act and rules relate to environmental protection:

- The Environment (Protection) Act, 1986
- The Environment (Protection) Rules, 1986

The Act was authorized to "accommodate the security and change of condition and for issues associated therewith." This demonstration characterized condition which incorporates "water, air, and arrive" and the between relationship which exists among and between “water, air and land, and people, other living animals, plants, miniaturized scale life forms and property.” It additionally characterized a perilous substance as “any substance or arrangement which, by reason of its synthetic or physical-concoction properties, or dealing with, is subject to make hurt individuals, other living animals, plants, microorganisms, property or the environment.” This law enrolls general forces of the central government which incorporated “every such measure as it esteems important or practical with the end goal of ensuring and enhancing the nature of the environment and anticipating, controlling and subsiding ecological contamination.”

This law requires that all organizations must have some kind of a Spill Prevention Control and Countermeasures Plan (SPCC). Environment reviewing is required by this law began in 1993. This report is to be submitted to the SPCB. The law demonstrates that, punishment for contradiction of the demonstration might be deserving of detainment up to seven years or fine up to Rs. 100,000 (€2500).

3.3 *Eco-mark*

To expand purchaser awareness, the Government of India propelled the eco-labelling plan known as ‘Eco mark’ in 1991 for simple ID of condition agreeable items. The consideration of textiles under ECOMARK as the eco-label is bringing new rivalry for the household industries. As whoever procures the name will be the main players in the residential market campaign. Any item which is made utilized or discarded in a way that fundamentally lessens the mischief it would some way or another reason to the environment could be considered as an Eco-Friendly Product. The materials with different items likewise shape its basic piece. The models determined under the ECOMARK plot are all earth amicable. To get it, the industry needs to cross every one of the obstacles of ecological guidelines. The criteria take after a support to-grave approach, i.e. from raw material extraction, to assembling, and to transfer. The ‘Eco mark’ is granted to buyer merchandise which meet the predetermined natural criteria and the quality necessities of Indian principles. Any item with the Eco mark will be the privilege natural decision.

Objectives of Eco-mark

- To give a motivator to makers and merchants to manufacture antagonistic ecological effect of items.
- To remunerate authentic activities by organizations to lessen unfriendly natural effect of their items.
- To help customers to end up plainly naturally capable in their everyday lives by giving data to assess ecological factors in their buy choices.
- To urge residents to buy items that has less harmful natural effects.
- Ultimately to enhance the nature of the earth and to empower the maintainable administration of assets.

3.4 Impact of EU Environmental Standards

Different ecological controls and guidelines concerning textile and dress have excited tremendous concern around the world. A few nations have started to change and update their own enactment and directions, while others are currently doing research in the field. As a rule, the EU countries have generally contrasting natural models as per their national needs. The most powerful ones are the German Act of 1994 prohibiting azo colors and the eco-labeling models concerning materials, for example, Eco-Tex. These ecological models and necessities have some positive and negative effects on worldwide exchange textiles.

The German boycott had prompted a considerable effect on India's fare. India's fare of the items utilizing German-prohibited colors represents 38% of the aggregate fare of textiles and apparel. The real fares incorporate cotton articles of clothing and fabrics, silk pieces of clothing and fabrics, floor coverings, carpets and a few chemicals. With this new direction from Germany, India had responded effectively for the most part through legitimate means and overhauling the capacity of acclimating to this new circumstance.

4 Eco-friendly Textiles

It is notable that each customer's item affects environment. However a normal purchaser does not know which item has less or more effect than the other one. Any item, which is made, utilized or discarded in a way that essentially diminishes the damage it would some way or another reason to the environment, could be considered as Eco-friendly item. Gradually, customers in India are taking lead in provoking makers to receive clean advances to create Eco-friendly items.

The majority of the garments in our closets contain polyester, Nylon or Lycra. These cheap and simple care fibers are spinning into the textile industry's wonder arrangement. In any case, their produce makes contamination and they are difficult to recycle (with nylon taking 30–40 years to decompose). The textiles and apparel industry is a differing one, as much in the raw materials it utilizes as the procedures it utilizes. At each of the six phases regularly required to make an article of clothing, the negative effects on the environment are as various as they are changed. Spinning, weaving and mechanical produce undermine air quality. Coloring and printing devour huge measures of water and chemicals, and discharge various unstable specialists into the air that are especially harmful to our wellbeing.

The expression "Eco-friendly" is by all accounts showing up wherever nowadays alongside different equivalent words "Naturally Friendly", "Nature Friendly" and "Green" however there is some misconception about what, in the event that anything, such terms really mean. The most straightforward meaning of Eco-friendly is that it portrays something that is naturally gainful or at any rate not hurtful.

4.1 Need for Eco-textile

- Need to put stock in ecological manageability making incremental strides in store network procedures to get it going.
- Need to know about the social, monetary and biological advantages of environment feasible item advancement forms.

4.2 Sustainable Fibers

Many individuals think about that as a 'sustainable fiber' is a natural fiber or a characteristic one. They will dismiss any man-made fibers on the ground that they harm the earth. However, some artificial or manufactured fibers can be more economical than characteristic ones as they do not use the same number of assets as the 'natural fibers'. The level headed discussion over how feasible natural fibers are situated all in all on the water and vitality utilization during the production of the fibers. Unless the fibers are natural, at that point unsafe chemicals are regularly utilized which harm the earth, as well as in charge of thousands of passing's a year. The water utilization of developing natural fibers frequently leaves others without clean water, and can harm the encompassing soil, making it infertile.

Ecofriendly fabrics are produced using fibers that don't require the utilization of any pesticides or chemicals to develop. They are normally impervious to shape and buildup and are without malady.

With respect to Man-Made Fibers (MMF), the MMF Industry specifically has a long history in the quest for activities which support sustainability. The dependable care program has been embraced in numerous nations, and has been connected by man-made fiber makers, since 1992. The utilization of vitality, raw materials and every single other resource, and furthermore the emanation of strong, fluid and vaporous waste decide the supportability of man-made fibers, much the same as for some other product. We should not overlook the being used stage, in which significant ecological reserve funds can be made, and the transfer or reusing stage. Enhancing the manageability of man-made fibers is the controlling standard to enhance ecological, economic and social execution.

4.3 Why Go Organic or Ecofriendly?

- **Social duty**: Chemicals and pesticides attack drinking water and groundwater, dirtying its fish and notwithstanding achieving human utilization. Natural and eco fibers develop with no pesticides or substance composts.

- **Biodegradable**: Eco and natural fabric biodegrade normally after some time. Manufactured fibers in the long run wind up plainly waste and let off destructive pollutant when they debase.
- **Health**: Many individuals are unfavorably susceptible or disdain wearing manufactured materials. Eco fabrics have every one of the properties of the new engineered breathable fibers with included delicate quality and wrap. They feel better against the skin.
- **Absorption**: Not just do its chemicals industry into the groundwater, customary attire is worn by the most permeable organ skin. Organic and Eco fibers are normal and don't contain chafing chemicals. Huge numbers of them are additionally considered hypoallergenic and normally anti-bacterial.
- **Popularity**: Organic nourishments have been around for a short time and it is a characteristic advancement that natural and ecofriendly fabrics will likewise pick up ubiquity. Eco and Organic fabrics once considered an option are presently going into the standard.

Common fiber, any hair like raw material specifically realistic from a animals, vegetable, or mineral source and convertible into nonwoven fabrics, for example, felt or paper or, subsequent to spinning into yarns, into woven fabric. A characteristic fiber might be additionally characterized as an agglomeration of cells in which the breadth is insignificant in examination with the length. Despite the fact that nature possesses large amounts of sinewy materials, particularly cellulosic types, for example, cotton, wood, grains, and straw, just a modest number can be utilized for textile products or other mechanical purposes. Aside from monetary contemplations, the value of a fiber for business designs is dictated by such properties as length, quality, malleability, flexibility, scraped area protection, retentiveness, and different surface properties. Most material fibers are thin, adaptable, and generally solid. They are versatile in that they extend when put under pressure and after that somewhat or totally come back to their unique length when the strain is expelled.

The utilization of common fibers for textile materials started before written history. The most established sign of fiber utilize is likely the disclosure of flax and fleece fabrics at unearthing centrals of the Swiss lake occupants (seventh and sixth hundreds of years BC). A few vegetable fibers were additionally utilized by ancient people groups. Hemp, apparently the most seasoned developed fiber plant, started in Southeast Asia, at that point spread to China, where reports of development date to 4500 BC. The craft of weaving and spinning cloth was at that point all around produced in Egypt by 3400 BC, demonstrating that flax was developed at some point before that date. Reports of the spinning of cotton in India go back to 3000 BC.

With enhanced transportation and correspondence, exceptionally confined abilities and expressions associated with textile produce spread to different nations and were adjusted to nearby needs and capacities. New fiber plants were additionally found and their utilization investigated. In the eighteenth and nineteenth hundreds of years, the Industrial Revolution energized the further production of machines for use in preparing different normal fibers, bringing about a gigantic upsurge in fiber production. The presentation of recovered cellulosic fibers (fibers framed of cellulose textile that has been broken down, decontaminated, and expelled), for example, rayon, trailed by the innovation of totally engineered fibers, for example, nylon, tested the imposing industry model of characteristic fibers for material and mechanical utilize. An assortment of engineered fibers having particular alluring properties started to infiltrate and overwhelm advertises beforehand hoarded by regular fibers. Acknowledgment of the focused danger from manufactured fibers brought about concentrated research coordinated toward the reproducing of new and better strains of common fiber sources with higher yields, enhanced creation and preparing strategies, and alteration of fiber yarn or fabric properties. The significant upgrades accomplished have allowed expanded aggregate generation, albeit characteristic fibers' real offer of the market has diminished with the convergence of the less expensive, manufactured fibers requiring less worker hours for production.

4.4 *Use of Eco Textiles Fibers—Related Enterprises*

- Fashion and Apparel Industry
- Home Furnishing and Textile Industry
- Hygiene and Health Care Industry
- Packaging Industry—"Eco Packaging" an Important Feature
- Growing Recycling Industry—Generating Rural Employments
- Medical Textiles Industry—Growing Opportunities

5 Qualities of Eco-friendly Fibers

- Eco well-disposed fibers are favored in hot and muggy atmospheres since they keep the body cool.
- They are biodegradable and have no negative impact on the environment.
- They are impervious to shape and mold and are sans sickness.
- They are developed without the utilization of pesticides and chemicals.
- Most of them are antibacterial, skin amicable and have certain recuperating properties.
- They help control different skin sicknesses caused by engineered and synthetically treated fabrics before.

The item was delivered without the utilization of brutal chemicals and pesticides, and isn't solid for the environment but on the other hand is health for you. Every single common fiber is not Eco-friendly. On the off chance that there is broad use of unsafe pesticides or bug sprays while developing, say, cotton-it doesn't remain Eco-friendly. Likewise the textile chemicals connected during the completing and different procedures of material generation obliterates the Eco-friendly character of these fibers. Nonetheless, in the event that it is a natural development with no pesticides and so forth and on the off chance that they are prepared with mechanical or different routes with no chemicals then just a fabric can be named as Eco-friendly. The same is valid for animals fibers like fleece where pesticides are utilized as a part of sheep plunges and hurtful medications are given to cure animal's infections however then natural fleece is there which an Eco-friendly fiber is moreover.

To the extent synthetic fibers are concerned they also can be Eco-friendly. These are recovered fibers. They are of two types—having protein inception and cellulose cause. Protein birthplace recovered fibers are extracted from plant protein like corn, soy, shelled nut and so forth or from animal protein like casein from drain. Recovered fibers of cellulose source are extracted from cellulose of wood pulp or leaves, for example, rayon fiber. On the off chance that fabricated in an earth neighborly way, they can be extremely well put into the classification of Eco-friendly.

6 Classification of Eco-friendly Fibres

The Classification of Eco-Friendly fibers are shown in Fig. 3.

6.1 Animal Fiber

Animal fibers are regular fibers that comprise generally of specific proteins. Examples are silk, hair/hide (counting fleece) and plumes. The animal fibers utilized most regularly both in the assembling scene and in addition by the hand spinners are fleece from central sheep and silk. Likewise exceptionally well known are alpaca fiber and mohair from Angora goats. Bizarre fibers, for example, Angora fleece from rabbits and Chiengora from canines likewise exist, however are once in a while utilized for large scale manufacturing.

Not every animal fiber have similar properties, and even inside an animal categories the fiber isn't reliable. Merino is a delicate, fine fleece, while Cotswold is coarser, but then both merino and Cotswold are types of sheep. This correlation can be preceded on the tiny level, looking at the measurement and structure of the fiber. With animals fibers, and common fibers all in all, the individual fibers appear to be unique, while every engineered fiber appear to be identical. This gives a simple approach to separate amongst normal and manufactured fibers under a magnifying lens.

6.1.1 Silk

Silk is a characteristic protein fiber, a few types of which can be woven into materials. The protein fiber of silk is made fundamentally out of fibroin and is delivered by certain creepy crawly hatchlings to shape casings. The best-known silk is extracted

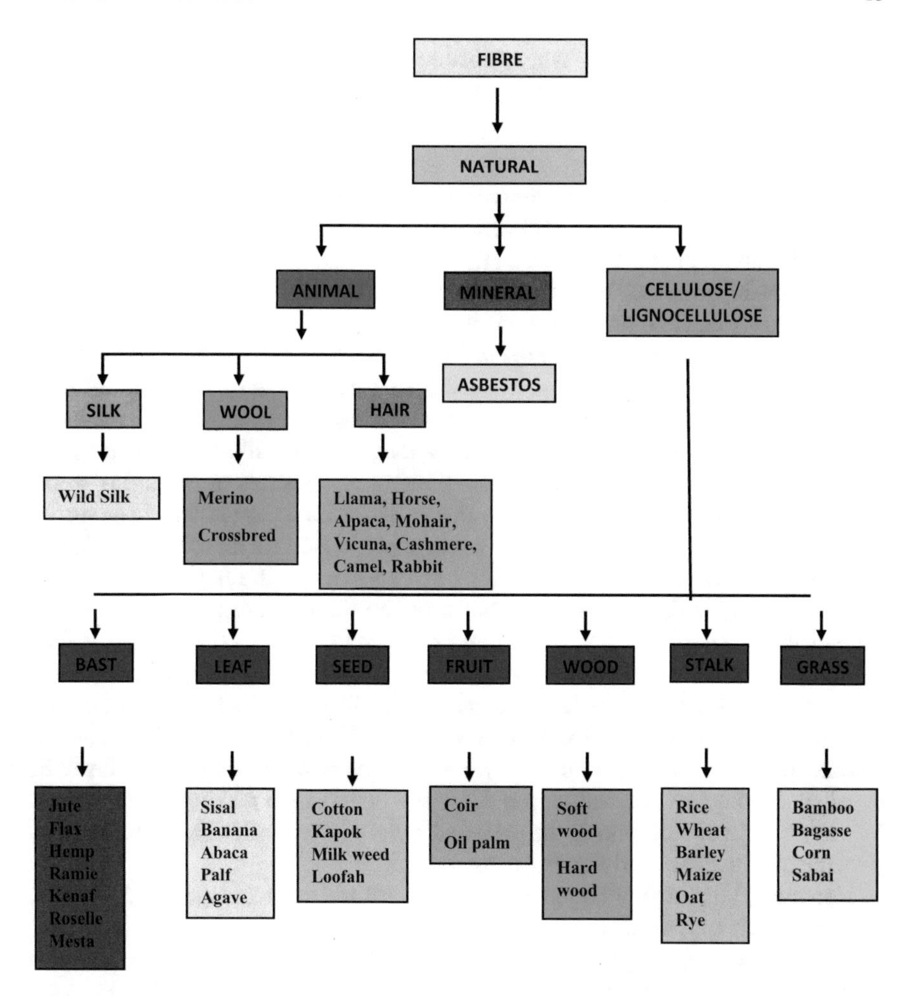

Fig. 3 Classification of fibers

from the covers of the hatchlings of the mulberry silkworm Bombyxmori raised in imprisonment (sericulture). The shining appearance of silk is because of the triangular crystal like structure of the silk fiber, which enables silk fabric to refract approaching light at various points, hence delivering diverse hues.

Silk is delivered by a few several insects, yet by and large just the silk of moth caterpillars has been utilized for textile assembling. There has been some exploration into different types of silk, which vary at the sub-atomic level. Silk is basically produced by the hatchlings of bugs experiencing complete transformation; however a few creepy crawlies, for example, web spinners and rough crickets deliver silk for the duration of their lives. Silk production additionally happens in Hymenoptera (honey bees, wasps, and ants), silverfish, mayflies, thrips, leafhoppers, scarabs, lacewings, bugs, flies, and midges. Other types of arthropod create silk, most prominently different 8-legged animals, for example, creepy crawlies.

Wild Silk

A few types of wild silk, which are created by caterpillars other than the mulberry silkworm, have been known and utilized as a part of China, South Asia, and Europe since old circumstances. In any case, the size of creation was constantly far littler than for developed silks. There are a few explanations behind this: in the first place, they contrast from the trained assortments in shading and surface and are thusly less uniform; second, covers accumulated in the wild have typically had the pupa rise up out of them before being found so the silk string that makes up the case has been attacked shorter lengths; and third, numerous wild casings are shrouded in a mineral layer that keeps endeavors to reel from them long fibers of silk. In this manner, the best way to acquire silk reasonable for spinning into textiles in territories where industry silks are not developed was by dreary and work concentrated checking.

Industry silks begin from raised silkworm pupae, which are reproduced to deliver a white-hued silk string with no mineral at first glance. The pupae are executed by either plunging them in bubbling water before the grown-up moths develop or by puncturing them with a needle. These elements all add to the capacity of the entire case to be disentangled as one ceaseless string, allowing a significantly more grounded fabric to be woven from the silk. Wild silks likewise have a tendency to be harder to color than silk from the developed silkworm. A procedure known as demineralizing permits the mineral layer around the cover of wild silk moths to be expelled, leaving just fluctuation in shading as a boundary to making a business silk industry in light of wild silks in the parts of the world where wild silk moths flourish, for example, in Africa and South America.

Hereditary alteration of tamed silkworms is utilized to encourage the production of more valuable types of silk.

India is the second biggest maker of silk on the world after China. Around 97% of the raw silk originates from five Indian states, to be specific, Andhra Pradesh, Karnataka, Jammu and Kashmir, Tamil Nadu and West Bengal. North Bangalore, the forthcoming site of a $20 million "Silk City" Ramanagara and Mysore, add to a greater part of silk produced in Karnataka.

In Tamil Nadu, mulberry development is gathered in the Coimbatore, Erode, Tiruppur, Salem and Dharmapuri regions. Hyderabad, Andhra Pradesh, and Gobichettipalayam, Tamil Nadu, were the primary areas to have mechanized silk reeling units in India.

India is likewise the biggest buyer of silk on the world. The convention of wearing silk sarees for relational unions and different propitious functions is a custom in Assam and southern parts of India. Silk is thought to be an image of sovereignty, and, truly, silk was utilized principally by the upper classes. Silk products of clothing and sarees delivered in Kanchipuram, Pochampally, Dharmavaram, Mysore, Arani in the south, Banaras in the north, Bhagalpur and Murshidabad in the east are all around perceived. In the northeastern province of Assam, three unique types of silk are delivered, on the whole called Assam silk: Muga, Eri and Pat silk. Muga, the brilliant silk, and Eri are delivered by silkworms that are central just to Assam.

6.1.2 Wool

Fleece is the textile fiber got from sheep and different animals, including cashmere and mohair from goats, qiviut from musk oxen, angora from rabbits, and different types of fleece from camel. Fleece essentially comprises of protein together with a couple of percent lipids.

A lot of the estimation of woolen textiles was in the coloring and completing of the woven item. In each of the focuses of the material exchange, the assembling procedure came to be subdivided into a gathering of exchanges, directed by a business visionary in a framework called by the English the "putting-out" framework, or "house industry", and the Verlags system by the Germans.

Super wash fleece (or launder able fleece) innovation initially showed up in the mid-1970s to deliver fleece that has been uncommonly treated so it is machine launder able and might be tumble-dried. This fleece is delivered utilizing a corrosive shower that expels the "scales" from the fiber, or by covering the fiber with a polymer that keeps the scales from joining to each other and causing shrinkage. This procedure brings about a fiber that holds life span and toughness over engineered textiles, while holding its shape.

Alpaca

Alpaca is the common fiber harvested from an alpaca. It is light or substantial in weight, contingent upon how it is spun. It is a delicate, strong, lavish and satiny natural fiber. While like sheep's fleece, it is hotter, not thorny, and has no lanolin, which makes it hypoallergenic. Alpaca is normally water-repellent and hard to touch off. Huacaya, an alpaca that develops delicate supple fiber, has normal pleat, consequently influencing a normally versatile yarn to appropriate for weaving.

Angora

Angora hair or Angora fiber refers to the fleece coat created by the Angora rabbit. While their names are comparable, Angora fiber is unmistakable from mohair, which originates from the Angora goat. Angora fiber is additionally particular from cashmere, which originates from the Cashmere goat. Angora is known for its delicateness, thin fibers. It is additionally known for its satiny surface. It is significantly hotter and lighter than fleece because of the hollow center of the angora fiber. It likewise gives them their trademark drifting feel.

Cashmere Fleece

Cashmere fleece, as a rule just known as cashmere, is a fiber acquired from cashmere goats and different types of goat. Regular use characterizes the fiber as fleece yet it is better and gentler than sheep's fleece. Some say it is hair, yet as observed underneath, cashmere requires the expulsion of hair from the fleece. The word cashmere is an old spelling of the Kashmir northernmost geological district of South Asia. Cashmere is better, more grounded, lighter, milder, and around three times more protecting than sheep fleece.

Mohair

Mohair is generally a silk-like fabric or yarn produced using the hair of the Angora goat. Both strong and flexible, mohair is eminent for its high radiance and sheen, which has helped pick up it the moniker the "Precious stone Fiber", and is frequently utilized as a part of fiber mixes to add these qualities to a textile. Mohair takes color outstandingly well. Mohair is warm in winter as it has astounding protecting properties, while staying cool in summer because of its dampness wicking properties. It is durable, normally flexible, fire safe and wrinkle safe. It is to be an extravagance fiber, similar to cashmere, angora and silk, and is generally more costly than most fleeces that originate from sheep.

Mohair is made for the most part out of keratin, a protein found in the hair, fleece, horns and skin of all mammals. While it has scales like fleece, the scales are not completely grown, just demonstrated. Consequently, mohair does not felt as fleece does.

Mohair fiber is roughly 25–45 μm in distance across. Fine hair from more younger animals is utilized for better applications, for example, garments, and the thicker hair from more seasoned animals is all the more regularly utilized for rugs and overwhelming fabrics proposed for outerwear.

6.2 *Mineral Fiber*

Various fibers exist that are extracted from characteristic mineral sources or are fabricated from inorganic and mineral salts. These fibers are transcendently subsidiaries of silica (SiO_2) or other metal oxides. Furthermore, metal fibers (either alone or typified in a reasonable natural polymer) are created. The regular element of these fibers is their inorganic or metallic composition and propensity to be warm safe and nonflammable, except for polymer-covered metallic fibers.

Asbestos is the name given to a few normal minerals (anthophyllite, amphibole, serpentine) which happen in a sinewy crystalline shape. The asbestos is at first soaked to open up the fiber mass, trailed via checking and spinning to yield fibers of round about cross area 1–30 cm long. Asbestos is extremely impervious to warmth and consuming, to acids and alkalies, and to different chemicals. In spite of the fact that it has low quality, asbestos fiber does not disintegrate in typical use, and it isn't assaulted by bugs or microorganisms. Asbestos is utilized as a part of flame resistant garments, transport lines, brake linings, gaskets, modern packing's, electrical windings, protections, and health proofing materials. Breathed in asbestos fibers have been appeared to be a genuine wellbeing peril, and it has been expelled from the materials showcase.

6.3 *Cellulose/Lingo Cellulose*

6.3.1 Bast Fiber

Bast fibers happen in the phloem or bark of specific plants. The bast fibers are as groups or fibers that go as strengthening components and help the plant to stay erect. The plants are gathered and the fibers of bast fibers are discharged from remaining of the tissue by retting, regular for disengagement of most bast fibers. The retted material is then additionally handled by breaking, scotching, and hackling.

Jute

Jute fiber is extracted from two herbaceous yearly plants, *Corchorus capsularis* (linden family, Tiliaceae) starting from Asia, and *Corchorus olitorius* originating from Africa. The former has a round seed unit, and the latter has a long case. Jute is developed principally in India, Bangladesh, Thailand, and Nepal. The plants are gathered by hand, dried in the field for defoliation, and water (pool) retted for periods up to a month. The profundity of the retting pools is reliant on the volume of precipitation during the storm season in Southeast Asia. Subsequently a year with less precipitation brings about low water levels in the retting pools and a lower review jute item because of defilement with sand and sediment. The fibers for export are evaluated for shading, length, fineness, quality, cleanliness, radiance, delicateness, and uniformity. The shading ranges from cream white to reddish brown, however more often than not the fiber has a brilliant shine. The fibers are polygonal in cross-segment with a wide lumen. Jute has customarily been an imperative material fiber, second just to cotton; notwithstanding, jute has been relentlessly supplanted by synthetics in the customary high volume uses, for example, cover sponsorships and burlap (hessian) fabrics and sacks. The fibers are likewise utilized for twine, while kraft pulping of jute gives extreme fibers for cigarette papers. The Indian government in collaboration with the United Nations Development Program has been engaged with a critical jute broadening system to discover new uses for jute in better yarns and materials, composites and sheets, and paper items. An especially encouraging outlet for jute is in formed composites with thermoplastic materials for inside car head, entryway and trunk liners.

Flax

The flax fiber from the yearly plant *Linum usitatissimum* (flax family, Linaceae) has been utilized since old circumstances as the fiber for cloth. The plant develops in mild, respectably wet atmospheres, for instance, in Belgium, France, Ireland, Italy, and Russia. The plant is additionally developed for its seed, from which linseed oil is produced. Another of the seed plant is the tow fiber utilized as a part of papermaking. The bast fibers are dew or water retted with dew retting for the most part yielding a dark fiber. Flax fiber is with high quality produced by water (stream) retting in the waterway of the river Lys in Belgium. The boiling, faded fiber contains nearly 100% cellulose. The flax fiber is the most strongest of the vegetable fibers, considerably more grounded than cotton. The fiber is very permeable, an essential property for garments, however is especially inextensible. The most essential application is in material for garments, fabrics, trim, and sheeting. Flax fiber is additionally utilized as a part of canvas, strings and twines, and certain mechanical applications, for example, fire hoses. Synthetic pulping of flax gives the raw material to creation of currency and composing paper. Flax fiber is likewise regularly utilized as a part of cigarette papers. Flax fibers are evaluated for fineness, delicate quality, extend, thickness, shading, consistency, shine, length, handle, and cleanliness.

Hemp

The hemp fiber is the plant *Cannabis sativa* (mulberry family, Moraceae) locating in central China. It is developed in central Asia and Eastern Europe. The stem is utilized for fiber, the seeds for oil, and leaves and flowers for drugs. The stalks grow 5–7 m tall and 6–16 mm thick. The hallow stems, smooth until the unpleasant foliage at the best, are hand cut and spread on the ground for dew retting for the most elevated quality item. Water retting is utilized on sun-dried groups from which the seeds and leaves have been evacuated. Fibers of hemp fiber can be 2 m long. The fibers are evaluated for shading, shine, spinning quality, thickness, cleanliness, and quality. It has a Z twist as opposed to the S twist of flax. Hemp is viewed as a substitute for flax in yarn and twine. Its prior use in ropes has been supplanted by leaf and engineered fibers. Hemp fiber is utilized as a part of Japan, China, Hungry, and Italy to make claim to fame papers, including cigarette paper, however bleaching is troublesome. The fiber is coarser and has less flexibility than flax. There is at present enthusiasm for reintroducing hemp into the United States and Canada as another fiber for farmers. In any case, this is fashioned with political and lawful issues because of the failure to recognize industrial hemp from hemp plants with high opiate content.

Ramie

The ramie fiber is situated in the bark of Boehmerianivea, an individual from the nettle family (Urticaceae). The plant is a native of China (subsequently, its name China grass), where it has been utilized for fabrics and fishing nets for a long time. It is likewise grown in the Philippines, Japan, Brazil, and Europe. The ramie plant grows 1–2.5 m high with stems 8–16 mm thick. The roots send up shoots on collecting, and two to four cuttings are conceivable every year, contingent upon soil and atmosphere. The plant is reaped by hand sickle and, after defoliation, is stripped and scratched by hand or machine decorticated. As a result of the high gum (xylan and araban) substance of up to 35%, retting is not conceivable.

The degummed, balanced fiber contains 96–98% cellulose. The ramie fibers are oval-like in cross-segment with thick cell walls and a fine lumen. The cell wall constituents in the ramie fiber, as other bast fibers aside from flax, have a counterclockwise twist. Ramie is the longest of the vegetable fibers and has astounding cluster and extraordinary quality; in any case, it has a tendency to be hardened and weak. Wet quality is high and the fiber dries quickly, benefits in fishing nets.

Traditional utilizations for ramie have been for substantial modern type fabrics, for example, canvas, bundling material, and upholstery. Expanded creation of the fiber in Asia, especially China, has advanced the utilization in mixed fabrics with silk, material, and cotton which would now be able to be found available.

Kenaf and Roselle

These firmly related bast fibers are extracted from *Hibiscus cannabinus* (mallow family, Malvaceae), individually. The fibers have other common names. Kenaf is produced for generation in China, Egypt, and regions of the previous Soviet Union; Roselle is produced in India and Thailand. Manor developed kenaf is fit for developing from seedlings to 5 m at development in five months. It is accounted for to yield ~6–10 ton of dry issue per acre of land, nine times the yield of wood. The plants are hand-cut, cut, or pulled in developing nations while motorized reaping techniques are under scrutiny in the United States. Roving machines are once in a while used to isolate the fiber-containing bark before retting for improving of the kenaf fibers. For pulping, the kenaf is fragmented or hammer milled to 5-cm pieces, washed, and screened. Kenaf fibers are shorter and coarser than those of jute. Kenaf fiber is likewise viewed as a substitute for jute and utilized as a part of sacking, rope, twine, packs, and as papermaking pulp in India, Thailand, and the Balkan nations. Roselle blanched pulp is vended in Thailand.

Urena

These are less imperative vegetable fibers of a jutelike nature. *Urena lobata* (Cadillo) of the mallow family (Malvaceae) is an everlasting plant that grows in Zaire and Brazil to a tallness of 4–5 m with stems 10–18 mm in diameter. On account of a lignified base, the stems are cut 20 cm over the ground. The plants are defoliated in the field and retted comparatively to jute and kenaf. The retted material is stripped and washed and again, rubbed by hand. The smooth, close white fiber is reviewed for gloss, shading, regulating, quality, and cleanliness. It is utilized for sacking, cordage, and coarse materials.

Abutilon Theophrastus

Abutilon Theophrastus is an herbaceous yearly plant producing a jute-like fiber. The plant is native to China and is monetarily grown in China and the previous Soviet Union. As a result of its relationship with jute in blends and fare, it is likewise called China jute. The plant develops to a stature of 3–6 in with a stem width of 6–16 mm. In the wake of reaping by hand and defoliation, packs of the stems are water retted and the fiber is separated by strategies like those for jute. The fiber is utilized for twine and ropes.

6.3.2 Leaf Fiber

Leaf fiber, hard, coarse fiber acquired from leaves of monocotyledonous plants (blooming plants that typically have parallel-veined leaves, for example, grasses, lilies, orchids, and palms), utilized mostly for cordage. Such fibers, normally long and solid, are likewise called "hard" fibers, recognizing them from soft and more adaptable fibers of the bast, or "delicate," fiber gathering.

Leaf fiber is for the most part acquired from sword-molded leaves that are thick, meaty, and regularly hard-surfaced, for example, those of plants of the agave family (Agavaceae), a major source. The leaves are reinforced and bolstered by fiber groups, regularly a few feet since quite a while ago, made out of many covering cells, or genuine plant fibers, held together by sticky substances. The fiber for the most part crosses the length of the leaf and is regularly densest close to the leaf undersurface. Leaves of the abaca plant, with fiber packs amassed in the stalks, are a special case.

The leaves are hand-harvested, and their fiber is isolated from the encompassing leaf tissue by decortications, a hand or machine scratching or peeling process, at that point cleaned and dried. The discharged fiber packages, or fibers, are not isolated into singular fiber cells and are called fibers in the exchange.

Leaf fibers are mainly utilized for such cordage as rope and twine. They may likewise be utilized for woven fabrics, more often than not requiring no spinning for this reason. Sisal, abaca, and henequen lead in world production. Numerous possibly helpful leaf fibers stay unexploited in view of the constraints of existing development and handling techniques and the expanded utilization of manufactured fibers for cordage. Economically useful leaf fibers incorporate abaca, cantala, henequen, Mauritius hemp, phormium, and sisal.

Sisal

The genuine sisal fiber from *Agave sisalana* is the most imperative of the leaf fibers as far as quality and commercial utilize. Beginning in the tropical western side of the equator, sisal has been transplanted to East Africa, Indonesia, and the Philippines. It is named after the port in the Yucatan from which it was first sent out. The sisal plants leaves grow from a central bud and are 0.6–2 m long, finishing in a thistle like tip. The fibers are emended longitudinally in the leaves, which are squashed, scratched, washed, and dried. The most elevated evaluations are additionally cleaned with a spinning drum. The development of the sisal plant relies upon water accessibility; it stores water during the wet season and expends it during times of dry spell.

The sisal fiber is coarse and solid, however contrasted and the abaca fiber it is inflexible, in spite of the fact that with a moderately high extension under anxiety. It is likewise impervious to salt water. The principal applications are in folio and baling twine and as a raw material for pulp for items requiring high quality. A substantial pulp process is working in Brazil in view of sisal.

Abaca

The abaca fiber is extracted from the leaves of the banana-like plant (same type) Musa textiles (banana family, Musaceae). The fiber is additionally called Manila hemp from the port of its first shipment, despite the fact that it has no association with hemp, a bast fiber. The well-develop plant has 12–20 stalks developing from its rhizome root framework; the stalks are 2.6–6.7 m tall and 10–20 cm thick at the base. The stalk has leaf sheaths that venture into leaves 1–2.5 m long, 10–20 cm wide, and

10 mm thick at the middle, the fibers are in the furthest layer. The plant produces a crop following 5 years, and 2–4 stalks can be collected about like six months.

Abaca fiber is one of a kind in its protection from water, particularly salt water, and it is utilized for marine ropes and links, in spite of the fact that it has been generally supplanted by synthetic fibers. Abaca fiber is the most strongest of the leaf fibers, trailed by sisal, phormium, and henequen; it is additionally the most strongest among the papermaking fibers. It is utilized for frankfurter housings and it is the favored fiber for tea packs as a result of its high wet quality, cleanliness, and structure that grants fast dispersion of the tea separate.

Caroa

Caroa is a hard leaf fiber, looking like sisal, acquired from *Neoglaziovia variegata*, a plant of the pineapple family (Bromeliaceae) developing wild in eastern and northern Brazil. The sword shaped formed leaves are 1–3 m long and 2.5–5 cm. wide. The fiber is separated by hand scratching subsequent to beating or retting. The fiber is utilized for cordage and acoustic material.

Henequen

Henequen fibers are white to yellowish red and are mediocre compared to sisal in quality, cleanliness, surface, and length, the other reviewing criteria. The lower base leaves, which are up to 2 m long and 10–15 cm wide, are cut, machine decorticated, and cleaned. Henequen is developed for nearby use in Cuba (Cuban Sisal) and El Salvador. Twine, ropes, coarse floor coverings, and sacks are made industrially from henequen.

Mauritius Hemp

Mauritius hemp, additionally called piteira, is extracted from *Furcraea gigantea*, likewise an individual from the Agavaceae. The plant is generally grown in the island of Mauritius, but on the other hand is reaped in Brazil and other tropical nations. The leaves are longer and heavier than those of the agaves. The fiber is removed by mechanical decortications. It is more white, longer, and weaker than sisal fiber. In view of its shading it is utilized as a part of mixes.

Phormium

The *Phormium tenax* plant yields a long, light-shaded, hard fiber otherwise called New Zealand hemp or flax, despite the fact that it has none of the bast fiber qualities. The plant is an enduring of the Agavaceae with surrenders over to 4 m long and 10 cm wide. The fibers are recovered by mechanical decortications.

Sansevieria

This type of the Agavaceae is an enduring otherwise called bowstring hemp from its utilization in bow strings. The plant is originated in tropical Africa and Asia however is broadly developed, essentially as an ornamental plant. It is of minor significance as a fiber plant, despite the fact that the fiber is of high quality. The most astounding evaluation of *Sansevieria cylindrica* fiber is greenish yellow, smooth and fine, and contrasts and sisal in quality.

Agave Cantala

Agave cantala is an individual from the agave family (Agavaceae) that incorporates sisal. It was originated in Mexico and was transported to Indonesia and the Philippines, where it is currently created industrially. The plant develops in a wet, sticky soil. The fiber is separated in Indonesia mechanically by a decorticator and in the Philippines by retting in seawater and cleaning by hand or with a decorticator. The cantala fiber is lighter in shading than different agaves, and its quality relies upon its arrangement.

6.3.3 Seed Fiber

The seeds and fruits of plants are frequently joined to hairs or fibers or encased in a husk that might be stringy. These fibers are cellulosic based and of commercial significance, particularly cotton, the most essential common textile fiber.

Coir

This fiber, acquired from husks of the fruit of the coconut palm, *Cocos nucifera* (palm family, Arecaceae), is basically created in India and Sri Lanka. The natural fruits are broken by hand or machine and the fiber extricated from the broken husks from which the coconut has been expelled for the copra. The husks are retted in waterways, and the fiber isolated by hand beating with sticks or by a decortication machine. The fibers are washed, dried, and hackled, and utilized as a part of upholstery, cordage, fabrics, mats, and brushes.

Kapok

Kapok fiber is extracted from the seed units of the kapok tree, *Ceiba pentandra*, of the Kapok tree family (Bombacaceae) which is indigenous to Africa and Southeast Asia; the fiber is produced predominantly in Java. The tree develops to a stature of 35 m. The seeds are contained in capsules or pods that are picked and torn open with hammers. The floss is dried and the fiber is isolated by hand or mechanically. A non drying oil is produced from the seeds with properties like cottonseed oil. The fiber is exceedingly light, with a circular about cross section, thin walls, and a wide lumen. Kapok fibers are dampness resistance, light, strong, delicate, and fragile, however not reasonable for spinning. The conventional uses were in life coats, resting sacks, protection, and upholstery. However, synthetics have supplanted the majority of the applications as filling material and now kapok is principally utilized forever preservers. Kapok-filled life preservers can support off to three times the heaviness of the preserver and do not end up plainly waterlogged.

7 Other Fibers

Sugarcane Fiber

The sugarcane fiber is extracted from the stick stalk of the organic product, otherwise called bagasse. The garment industry utilizes bagasse, basically a waste item, to make textile fibers. Lyocell is a sugarcane fiber produced using the bagasse having enormous potential application in the medicinal field. The constancy, non-abrasiveness, and the biodegradability of the fibers make them the ideal raw material for making dispensable therapeutic and commercial material. Sugarcane fibers utilized with a mix of selvage denim are additionally being utilized to make denim jeans and coats utilizing the well-established Japanese procedures.

Utilization of organic product fibers makes the textile materials earth well disposed, natural, and biodegradable stuff. In addition, since a few fibers are likewise created from side-effects and squanders of the plants, they are additionally recycled fibers. Natural products, which taste great as well as look great, are gradually advancing into the textile world.

Banana Fiber

Another new type of characteristic fiber is acquired from the banana plant. The bark of the banana tree is being utilized to make fibers. The normal fineness of the banana fiber is 2380 Nm and the standard length of the fiber is 60 mm. The banana fiber characterized of cellulose, lignin, and hemicelluloses. These fibers look basically the same as bamboo and ramie fibers but the smoothness of the banana fibers is much better. Bananas have not much scope of textile application. They were being utilized to fabricate items like ropes, mats, and other composite materials. Recently for Eco-friendly material fabrics, the significance of banana fibers is rising. One can discover pieces of clothing made out of the fibers alongside neck ties, pad covers, table fabrics, and shades. Physical properties of banana fibers like solid dampness retention, common sparkle, high quality, and little stretching make it perfect for clothing generation.

Pineapple Fiber

The much enjoyed natural product is currently being utilized for making textile materials. In any case, it is not the organic product however the leaves that are utilized to make the fiber. The fiber from the leaves is extracted either by manually retting the leaves submerged or by mechanically extricating them. The remaining parts of the pineapple contain high measure of lignin and cellulose. The fibers extricated from the fruit have silk like luster and have cream shading. They are better than jute, have anti- bacterial, and dyeing properties.

Blending pineapple fibers on account of its silk like qualities with polyester and silk have found new utilities in the textile industry. Also, highlights like high quality, biodegradability, and moderate evaluating influence the pineapple to leaf fiber a decent hotspot for assembling specialized materials as well. Fabrics produced using this delectable fruit product is delicate, measure light are anything but difficult to deal with and wash. Philippines are the biggest maker of pineapple fiber on the world. The fiber is utilized to fabricate upholsteries, home materials, and garment, non-woven, and specialized material fabrics.

Corn

Corn fiber has comparative qualities to polyester staple fiber and has the shine of silk, then its moisture recover outperform polyester, so the fabric made of it is much agreeable. Its adaptability and creep recovery is good so the fabric has great shape maintenance and hostile to wrinkle, it has incredible and touch and wrap, good dye ability and it can be dyed with dispersion dyes under typical pressure, and it has the excellent anti to fade in color and unaffected by UV light.

Bamboo Fiber

The bamboo woods in China have to a great extent been developed there for a large number, and now and again thousands, of years. Consistently, in spring and summer, new shafts (called culms), develop from a shoot underground. One shoot of the larger timber assortment of bamboo weighs between 2–4 kg when it is under 30 cm high. At this stage it is very delicate and can without much of a stretch be cut with a knife. In the event that left to develop, this shoot achieves its full stature of say 20 m in a mind blowing 3 months, (tallness relying upon assortment). When mature, the timber is uncommonly flexible and solid. It has an elasticity very gentle steel.

Bamboo textiles fiber is produced using bamboo timber which has developed in the timberland for no less than 4 years. Indeed, even in remote regions of China bamboo woodlands are exceedingly esteemed and painstakingly tended and oversaw. In summer, when new shoots achieve their full tallness, they are set apart with a year code which ensures they are collected at the correct development. At the point when collected they are taken to factories where they are squashed and submersed in an arrangement of sodium hydroxide which breaks up the bamboo cellulose. With the expansion of carbon disulfide it renders the blend prepared to recover fibers which are then drawn off, washed and blanched to a bright white shading and dried. The resultant lighten is long in staple and obviously better than different fibers. At that point they are spun into yarn, similar to some other textile fiber. The more drawn out staple and higher rigidity is the thing that makes an extreme, delicate yarn—which is not as defenseless to wearing and fraying the same number of different yarns. This is the thing that gives bamboo fabrics phenomenal toughness. The void of the fiber adds to its abnormal state of permeableness. It takes more time to dry on a clothesline. The void of the fiber likewise empowers it to hold colors and shades all the more promptly and for all time, hence making it substantially more colorfast.

Soya Fiber

Soybean is one of the nature's magnificent nutritious blessings. Soybean (Glycine max) is a leguminous plant. It is one of the very few plants that give a fantastic protein with least soaked fat. Soybean enables individuals to rest easy and live longer with an improved personal satisfaction. India is the fifth biggest producer of soybeans on the world. Soya Protein fiber (SBF) is a type of regenerative plant fiber. SPF is the main plant protein fiber on the world, a recently conceived protect to humanity's skin.

Mudar Fiber

Mudar fibers are cellulosic seed fibers developing on single cells in a large seed of plant developing on single cells in a large seed of the plant. Fibers are hollow with thin dividers in respect to their distance across, and are in this manner lightweight. Attributable to their structure, they are utilized as a part of things where good protection and lightness properties are required. Mudar fibers are described by high dampness recapture that can be emphatically used yet which frequently causes the fiber masses to end up plainly damp and clamp together. Mudar fiber (drain weed) has a place with the variety calotropis of the family asclepiadaceae.

The members from this family are mostly found in the tropical regions of the world. They happen particularly in the drier parts of central and South America. The plants of this family are for the most part erect herbs or woody climbers however some are succulent. *Calotropis gigantea* and *Calotropis procera* are a significant in squander lands and by the roadside, frequently on dark cotton soil. In the second stage the follicle operation winds up. Normally the seeds are flat, praise, elliptical, and are delegated by a fruit of hairs. These hairs encourage the dispersal of the seeds by wind. At the point when the fibers are dry, they discrete. The units are picked when still green and unopened. They are spilt and the green husk pulled back, which reveals within. The milkweed fibers, with their seeds appended, are expelled from the opened unit. At the point when the appended seeds are rubbed gently against the palm of the hand, they tumble off promptly from the fibers.

Milk Protein Fiber

Milk Yarn is produced using milk protein fibers. To manipulate it, milk is first dewatered, i.e. all the water content is taken out from it and after that it is skimmed. New bio-engineered technique is then applied to make a protein spinning liquid. This liquid is reasonable for wet spinning process through which the final high grade material fiber is made. While spinning, a dissolvable is utilized by a large portion of the producers and smaller scale zinc particle is implanted in the fiber which gives it the qualities of being bacteriostatic and strong. It consolidates the upsides of both, natural and also engineered fibers. Milk protein fiber is a type of new fiber that has exceptionally solid capacities. It contains eighteen amino-acids, which are valuable to human wellbeing. It can be spun alone or with cashmere, silk, spun silk, cotton, fleece, ramie and different fibers to make fabrics. The fabrics made of these fibers sustain and deal with skin in an exceptionally productive way by keeping without end sensitivities and even wrinkles. The fabrics produced using milk yarn are fundamentally utilized as a part of assembling kids wear, top-grade clothing, shirts, T shirts, loungewear, and etc.

Aloe Vera Fiber

Aloe Vera is fundamentally a central plant of Africa. It is otherwise called lily of the desert and plant of everlasting status because of its therapeutic impacts. This plant has 96% of water content. The leaf of this plant contains more than 75 supplements and 200 numerous, for example, 20 minerals, 18 amino acids and 12 vitamins. The substance of the plant helps in restoring the skin cells helps the arrangement of health dermis and battles against skin harm. Advancements in the material field have inserted the excellences of aloe vera in articles of clothing, which anticipates maturing of the skin; revives skin cells, and keeps skin free from microbial contaminations. Garments are made of a small scale fiber with an open construct that enhances the transport of dampness to the skin.

Nettle Fibers

Stinging nettle fibers can be utilized to make Eco-friendly apparel. Stinging nettle grown in the wild and is broadly known as weed. Nettle fiber can be spun into a yarn and utilized as a fabric for garments and numerous different uses. Stinging nettle is a characteristic moth repellant and nettle fiber has been utilized for attire in many cultures.

8 New Fibers in the Market

Merino Fleece

Merino fleece has spined into an inexorably well-known other option to engineered fabrics regularly utilized as a part of athletic and outside wear. The fleece is apparently the mildest on the world, and in light of the fact that the sheep grows winter coats each season, is not a normally sustainable asset. While considering the utilization of merino fleece for clothing, it's constantly vital to consider the treatment of the sheep and to ensure the regular routine with regards to 'mulesing' is not utilized by the agriculturists. Also, alpaca fleece, cashmere, mohair, sisal and abaca are viewed as low-affect natural fibers perfect for attire.

Peace Silk

Peace silk is a characteristic silk that does not require the boiling of live silkworms in their cases. Peace silk takes into consideration the worm to rise normally from its cocoon before the cocoon is spun into string, and in the end transformed into silk. Ventures to help peace silk are winding up progressively more well-known with craftsman being instructed the new, less intrusive generation process.

Eucalyptus Tree Fiber

Eucalyptus is the plant used to create Tencel® and is grown on minor land inadmissible for developing nourishment. These trees are grown with least water utilize while utilizing manageable ranger service activities. Since the Tencel® procedure is "shut circle," it's been considered the fiber without bounds. It begins with the extraction of raw material wood before a 100% "recoverable" dissolvable is connected to separate the wood pulp into fiber. The production procedure won the "European Award for the Environment" by the European Union.

Qmilk Fiber

Qmilk fibers are 100% natural, sleek and healthy skin. They meet the prerequisites for inventive material improvements. With a characteristic antibacterial impact and solid hydrophobicity, they give fibers to the additional estimation of the items in the development showcase.

Qmilk is the natural fiber to have thermo-holding properties. In this way, other common fibers can likewise be joined without ordinary plastics or phenolic gums. In this way, lightweight development stays 100% natural and can be treated the soil.

With the universes first—100% natural plastic, QMILK makes a conclusive commitment to the strength of individuals and surroundings.

9 Advantages and Benefits of Natural Fibers

Normal fibers have the following advantages

- Environmentally friendly
- Fully biodegradable
- Non-lethal
- Easy to handle
- Non grating during preparing and utilize
- Low thickness/light weight
- Compostable
- Source of income for rural/farming group
- Good protection against warmth and clamor
- Renewable, abundant and ceaseless supply of raw materials
- Low cost
- Enhanced vitality recuperation
- Free from health hazards (cause no skin aggravations)
- Acceptable particular strength properties
- High strength
- Good thermal properties
- Reduced tool wear
- Reduced dermal and respiratory irritation

- Ease of partition
- The grating nature of natural fibers is much lower contrasted with that of glass fibres, which offers advantage as for handling methods and recycling.

10 Textile and Apparel Industry-Go Green

The utilization of natural fabrics is not the main way the design business could go "green" and secure the eventual fate of the world and its regular assets. Organizations must be additionally more moral and utilize reasonable exchange and reasonable work; this is the "ethical and ecological mindfulness". This implies organizations are paying the reasonable cost, making reasonable business openings, creating safe workplaces for the specialists that they get their materials from. In addition, they are "participating in earth supportable works on, ensuring that product quality is kept up, respecting social way of life as a boost for item improvement and creation works on, offering business and specialized skill and opportunities for laborer progression, adding to group advancement, fabricating long term exchange connections, and being interested out in the open responsibility." In this unique situation, both reusing fashion and organic clothing can add to an eco-feasible advancement.

Organic

Textile, to qualify as being natural, should to be produced using characteristic sources, for example, plants or animals, gathered or natural fabricated. Colors utilized on natural apparel should to be either plant/mineral determined or ecologically low effect colors. No overwhelming metals or other destructive chemicals should to be utilized as a part of the dyeing procedure. A morally and ecologically mindful approach should to be stretched out to every one of the parts of the organic apparel industry.

Clean by Design

Clean by Design is an inventive program that uses the buying power of multinational companies as a lever to lessen the natural effects of their providers abroad. Clean by Design concentrates on enhancing process proficiency to decrease waste and emanations and enhance the environment. The impacts of the apparel Industry is shown in Fig. 4.

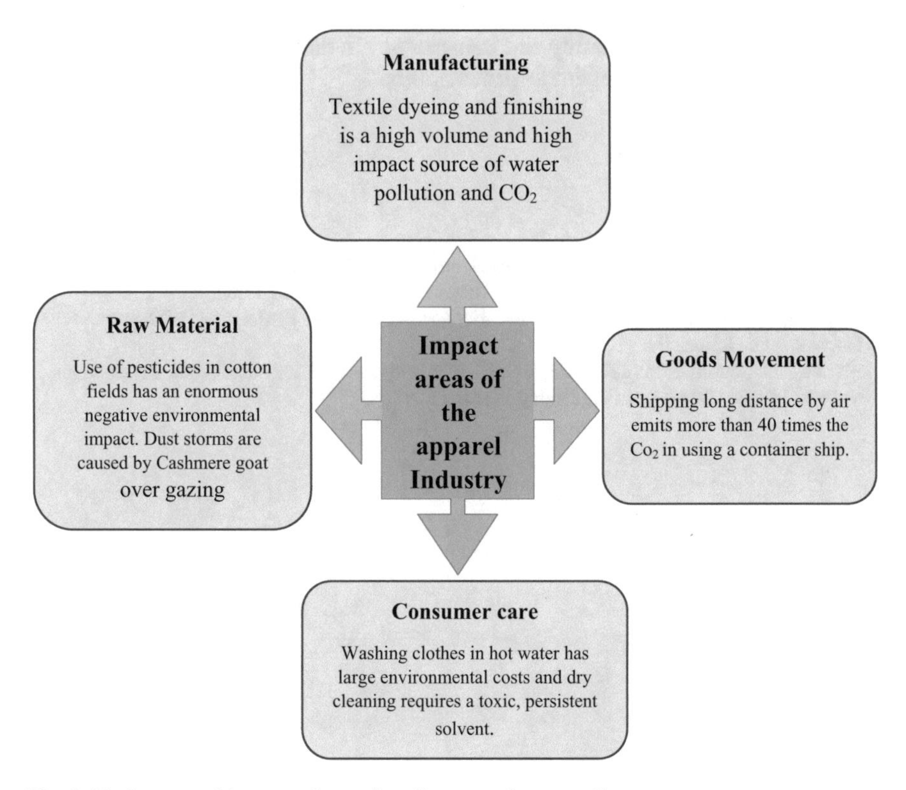

Fig. 4 Environmental impacts. *Source* http://www.environmentalissues.net

11 Conclusion

In current situation, the interest for Eco-textiles has kept on developing. To caring of this demand without yielding the human health and the world wellbeing, one must sustainable textile solutions. The fabrics made out of eco fibers can be worn by any one as they do not have any aggravating chemicals in them. Thus the use of eco fibers and natural are the best explication for keep our earth clean and to reduce global warming. The textile industry is also responsible for majority of the total industrial pollution that affects land and waters. In order to prevail over these difficulty and environmental collision, organic farming helps in the cultivation of highly famine-tolerant crops. They are developed to be eco-friendly, for waste-to-wear technology. The processing and supply chain is also concerned about the long-term health of the planet, by reducing the release of CO_2 into the environment. Even though Eco-Friendly fibres are getting used now days in many clothing applications, still there is a scope for them to completely replace non Eco-Friendly fibres in all of the clothing applications and many such Eco-Friendly fibres and eco textiles should completely reach every consumer. This chapter discusses the upgrading of technology that results in a less concentrated feed, no or a less harmful effect on the environment and traceability and transparency in the product, which guarantees minimum standards with regard to sustainability.

References

Blackburn RS (2005) Biodegradable and sustainable fibres. Woodhead Textile Series, USA

Fletcher K (2008) Sustainable fashion and textiles, 1st edn. Earthscan, London

Horrocks AR (1998) Ecotextile '98, sustainable development. In: Proceedings of the conference, University of Bolton, UK

Horrocks AR, Miraftab M (2007) Ecotextiles—the way forward for sustainable development in textiles, 1st edn. Wood head Publishing Limited

Regenerated Sustainable Fibres

Shanmugasundaram O. Lakshmanan and Guruprasad Raghavendran

Abstract In the present era of environmental consciousness, sustainable materials play a vital role in protecting public health and environment. The main problems with synthetic polymers are that they are non-degradable and non-renewable. In the last few decades, the textile industry has witnessed introduction of many new fibres in the market. These fibres are biodegradable and are extracted/manufactured from annually renewable crops and other sources. Many innovations have been reported in recent years regarding the sustainability of these fibres. In this chapter, the production, properties and applications of some of the sustainable fibres like Polylactic acid (PLA), Lyocell, Regenerated wool protein and Chitosan fibres have been discussed in detail. These fibres find application in variety of end-uses such as in apparel, home textile and medical textiles. Sustainable innovative measures in production and processing of these fibres are discussed.

Keywords Chitosan · Eco-friendly and sustainable fibres · Lyocell · Polylactic acid · Wool keratin

1 Introduction

The raw materials and their products have to be constantly environmental-friendly during the life cycle in order to protect the public health and environment. The eco-friendly fibres/fabrics not only create less negative impact on environment but also provide social and economic benefits. The International Organization for Standardization (ISO) standardized the life cycle assessment (LCA) of products over their entire life cycle; starting from production of raw material to final disposal to the environment. The synthetic fibre industry recorded a demand of 55.2 million tons in

S. O. Lakshmanan (✉)
K.S. Rangasamy College of Technology, Tiruchengode, Tamil Nadu, India
e-mail: mailols@yahoo.com

G. Raghavendran
ICAR-Central Institute for Research on Cotton Technology, Mumbai, India

S. S. Muthu (ed.), *Sustainable Innovations in Textile Fibres*, Textile Science and Clothing Technology, https://doi.org/10.1007/978-981-10-8578-9_2

the year 2014 (Textile world, 2015). The synthetic fibre consumption is huge globally with polyester, polypropylene and nylon dominating the market. These fibres are not biodegradable and cause big problems when the garment reaches the end of its cycle. Though recycling options are available which can solve the problems to some extent, much of the used synthetic garments are either incinerated or sent to landfill. Many scientific groups around the world are concentrating on development of bio-fibres to replace a part of synthetic fibres usage. The depletion of petroleum reserves is also a major reason for fibre industries to look for solutions. The natural fibres are environmental friendly but the amount of pesticides and chemicals used for the cultivation is a major concern for the environmentalists. The concern for the degrading environment conditions due to irresponsible use of chemical products have led to worldwide efforts to develop eco-friendly fibres with a vision to bring about a drastic reduction in global consumption of harmful non-biodegradable products. While replacing petrochemical products by bio based materials, it is essential that the quality and performance should not be affected to a major extent. Some of the consumer survey reports published indicated that consumer is not ready to compromise on the performance of an eco-friendly product (Thiry 2007; Anon 2007a). Many research works are being carried out to replace plastic, films, packing materials, building materials, and synthetic fibres in order to safeguard the environment. Due to the untiring efforts of researchers, many environment friendly fibres appeared on the textile scene like bamboo, modal, regenerated protein, casein, chitin, chitosan, alginate, polylactic-acid (PLA), lyocell etc. These fibres are used not only for apparel purpose but also various home textile, hygiene and biomedical applications. Many companies developed regenerated protein fibres such as casein (Courtanlds Ltd.), groundnut protein (ICI), soybean (Ford Motor Company), corn fibre (Virginia—Carolina Chemical Corporation), etc. The success of these eco-friendly fibres depend on their performance properties, cost, and sustainability aspects as well.

In this chapter, the production, properties and applications of some of the sustainable fibres like Polylactic acid (PLA), Lyocell, Regenerated wool protein and Chitosan fibres have been discussed in detail along with the recent sustainable innovations taken place in the production and processing of these fibres. Polylactic acid (PLA) is linear aliphatic thermoplastic polyester derived from 100% renewable resource such as corn. Lyocell fibres are manufactured from wood pulp produced from sustainable sources and use an eco-friendly solvent for its production. Keratin extracted from wool, feathers and horns has received good attention for application in biomedical fields due to their biocompatibility, cellviability and biodegradability (Yamouchi et al. 1998). Nanotechnology can be applied for the development of nanokeratin fibre that can lead to production of innovative green products with improved properties for applications in textiles, biomedical and biocomposites. Ionic liquids (environmental friendly solvents) are used to extract keratin from feather to increase dissolution of polymer. Chitin is a common bio product found in the exoskeleton of crustaceans and insects (spiders, shrimps and crabs). Chitosan is derived from chitin by deacetylation process. Innovative use of chitin and chitosan polymer in the form of nanofibers, membranes, scaffolds, beads, films, hydrogels, and sponges for biomedical applications were reported.

2 Polylactic Acid Fibres

2.1 Introduction

Polylactic acid (PLA) fibre is an aliphatic polyester derived from 100% renewable resources. PLA can be defined as a manufactured fiber in which the fiber-forming substance is composed of at least 85% by weight of lactic acid ester units derived from naturally occurring sugars. Polylactic acid or PLA is the first commodity polymer produced from annually renewable resources like corn, wheat starch and beet. PLA fibres can be melt spun or wet spun, which is an advantage. Extruded fibres can then be cut into staple fibres to blend with other staple fibres. The fibre possesses interesting properties which are useful for various end-use applications. Ingeo™ from Natureworks and ecodear™ from Toray industries are some of the popular brands of PLA fibres in the market.

2.2 Production Process

The monomer used to manufacture polylactic acid is obtained from annually renewable crops like corn, wheat starch and sugar beets. The plants are put through a milling process extracting the starch (glucose). Enzymes are added to convert the glucose to dextrose via a process called hydrolysis. The carbon and other elements in these natural sugars are then converted to lactic acid through fermentation. The polymer is then formed either by (1) direct condensation of lactic acid or (2) via the cyclic intermediate dimer (lactide), through a ring opening process. The direct condensation of lactic acid involves the removal of water by condensation and the use of solvent under high vacuum and temperature. Low to intermediate molecular weight polymers only can be produced by this route, mainly due to the difficulties of removing water and impurities. The Ring-opening polymerization is considered to be a better way to produce a high molecular weight polymer, and has now been adapted commercially. The process is based on removing water under milder conditions, without solvent, to produce a cyclic intermediate dimer, referred to as lactide. Ring-opening polymerization of the optically active types of lactide can yield a 'family' of polymers with a range of molecular weights by varying the amount and the sequence of D-lactide in the polymer backbone. Polymers with high L-lactide levels can be used to produce crystalline polymers while the higher D-lactide materials (>15%) are more amorphous. Based on this lactide intermediate method, NatureWorks LLC has developed a patented, low cost continuous process for the production of lactic acid based polymers. Many catalyst systems have been evaluated for the polymerization of lactide including complexes of aluminum, zinc, tin, and lanthanides. Metal alkoxides are the most common metal-containing species for the ring-opening polymerization of cyclic esters. In the process of polymerization, the lactide ring is opened and linked together to form the long chain of polylactide polymer (Farrington et al. 2005). The

Table 1 PLA Fibre properties

Property	PLA
Fineness (denier)	1.5–7
Fibre density (g/cm^3)	1.25
Tenacity (g/d)	2.5–5
Elongation (%)	10–70
Moisture regain (%)	0.4–0.6
Melting Point (°C)	160–170

long chain of PLA polymer is converted into pellet from and transported for transforming them into products including capsules, plastic cups, T-shirts, baby wipes, and appliances.

2.3 Polylactic Acid Fibre Properties

PLA fiber has a number of characteristics that are similar to many other thermoplastic fibers, such as controlled crimp, smooth surface and low moisture regain. The physical properties and structure have been studied by several researchers (Drumright et al. 2000), and these works confirmed that this polymer has significant commercial potential as a textile fiber. Its mechanical properties are considered to be broadly similar to those of conventional PET (Lunt and Bone 2001). The fibres are generally circular in cross-section and have a smooth surface. The properties of PLA fibre is given in Table 1.

The staple fibres are available in the market in different cut lengths of 38, 51 and 64 mm. The fibers are also available in dope dyed form. The tensile property of PLA fibre is different to that of high tenacity polyester and it is similar to that of wool. The fibres stretch easily after the yield point. Since the fibre extension is quite high, higher work of rupture is achieved ensuring acceptable performance in commercial use. The PLA fibre has a higher LOI (limiting oxygen index) of 26% compared to most other fibers, meaning that it is more difficult to ignite. The UV resistance of the fibre is also good. PLA shows good wicking behavior and because of faster spreading and quick drying, the moisture management capability of the fibre is excellent. As PLA is a linear aliphatic fiber, its resistance to hydrolysis is therefore relatively poor. Hence care must be taken in chemical treatments of the fibre. The Ingeo6 series of PLA fibres from Natureworks is designed for fiber processes from mono to multifilament as well as spunbond and meltblown products. The melting point ranges from 130 to 170 °C with amorphous to crystalline grades.

2.4 *Yarn and Fabric Production*

The production of filament yarns or production of staple yarns from PLA is fairly straight forward and do not cause more complexities in mechanical processing. The fibre more or less behaves like PET polyester. In the category of spun yarns, the commercial products range from Ne 5 to Ne 60 whereas in filament production, fineness of 70–150 dtex are common. The PLA material can be blended with other materials like cotton, viscose, polyester and wool easily. However the resultant properties of yarns depend on the blend composition. A combination of cotton with PLA results in moderate yarn tenacity, which is suitable for knitting applications only. But the positive attributes of PLA fibres like better moisture managing capability can be derived in the final product even with small quantity of PLA added with other fibres. The yarns are processable in knitting and weaving machine without any special requirements. In case of weaving, the yarns may be sized with PVA or water soluble size to avoid use of alkali for desizing. It was reported that PLA woven filament fabrics give a very soft hand, and have a high fluidity/drape compared to PET woven fabrics.

2.5 *Wet Processing of Polylactic Acid (PLA)*

The melting point of the PLA yarns that are commercially available today is relatively low at 170 °C. This causes problems in downstream textile processing, in particular on the chemical processes. PLA blended fabrics need to be scoured in mild alkaline conditions to avoid excessive weight loss (Parmar et al. 2014). The pre-treatment processes typically used for cotton are in the alkaline condition which can deteriorate the PLA fiber. Therefore, the pre-treatment conditions for the PLA/cotton blended fabric needed to be optimized to avoid fiber damage (Phatthalung et al. 2012). It was observed by Baig and Carr (2014) that during scouring PLA fabric was degraded at high alkali concentrations and processing temperatures. The scouring temperature above 60 °C causes damage to the fibre as solution penetrate the fibre structure and damages it. On treatment with hydrogen peroxide by the Cold Pad Batch technique the strength reduction was not appreciably high when compared to exhaust technique. Similar to PET, PLA is dyed with disperse dyes. However, dye selection is most important, as the individual dye behavior is quite different from dyeing on PET. In general terms, dyes show their maximum absorption at a shorter wavelength than on PET and tend to look brighter (Nakumara 2003).

2.6 *Garmenting and After-Care*

The low melting point of PLA fibre demand proper precautions to be taken during garmenting. As the cutting and sewing operations generate lot of fibre to metal friction which could generate heat, the material needs to be handled with more care. The final pressing stage is also important as high temperatures will deteriorate the material. Pressing operation has to be performed without heat or with mild heat only. The main concern lies in after-care of garments by consumer. The garment pressing and ironing temperatures have to be lower than other fibres like cotton and PET.

2.7 Applications of PLA

PLA fibres are useful in variety of applications in apparel and technical textile segments. The possible applications of this fibre in apparel include lingerie, men's shirts, active sportswear, outerwear, and performance wear. The positive attributes of PLA fibre can be made use of by blending PLA with other fibres like cotton, viscose, wool etc. Guruprasad et al. (2015) blended PLA with cotton fibres to improve moisture management properties of the blended material. It was reported that Overall Moisture management capability of the blended material was superior to 100% cotton material. The resistance to UV and low flammability, low smoke generation and low toxic gas on burning are attractive properties for the home textile market segment, where PLA can find application as bed sheets, pillow covers, curtains and drapes. The nonwoven produced from PLA fibres can be used for industrial/household wipes, diapers, feminine hygiene products and medical applications. The material is already much used in medical applications as suture material.

2.8 Recent Innovations

PLA fiber has been considered as a new polyester fiber for the textile industry with possible applications in apparel, home and technical textile applications. Ingeo6 series from Natureworks designed for fiber processes from mono to multifilament as well as spunbond and meltblown products. Melting point ranges from 130 to 170 °C with amorphous to crystalline grades. Toray is promoting the sale of ECODEAR™ PLA mainly as an industrial material, especially for automobile components, and also as a life material and green material. For the improved stake in apparel and technical segment, the fibre has much to do with issues like low thermal stability and extra care to be taken in garment pressing and ironing. The cost of the fibres is also relatively high compared to PET polyester. The development of fibres with improved thermal resistance will result in ease of processability and may open up new applications for this interesting fibre. As sustainability issues are gaining momentum nowadays, PLA seems to be a more promising material for future.

3 Lyocell Fibres

3.1 Introduction

The cellulose fibers produced by direct dissolution have the generic name of lyocell. Lyocell is the first in a new generation of cellulosic fibres made by a solvent spinning process. A major driving force to its development was the demand for an ecofriendly process that utilizes renewable resources as their raw materials (White et al. 2005). The so-called rayon process uses toxic chemicals to prepare the spinning solution. Therefore many attempts were made to invent new solvents to directly dissolve cellulose. Among these, *N*-methylmorpholine-*N*-oxide (NMMO) turned out to be the best solvent, leading to the commercial success of cellulose fibers under the trade name of Tencel by Courtaulds in 1994. Lenzing, Austria acquired the business of Tencel from Courtaulds, but continued with the brand name of Tencel. This cellulosic fibre is derived from wood pulp produced from sustainable sources. The wood pulp is dissolved in a solution of an 'amine oxide' (usually *N*-methylmorpholine-*N*-oxide). The solution is spun into fibres and the solvent extracted as the fibres pass through a washing process. The manufacturing process recovers >99.5% of the solvent. The solvent itself is non-toxic and all the effluent produced is non-hazardous.

Lyocell fibres offer wide range of properties that suits the requirements of apparel and home textiles. A range of attractive textile fabrics can be made from lyocell that are comfortable to wear and have good physical performance. The performance properties combined with absorbency has made lyocell ideal for nonwoven fabrics and papers. Lyocell has all the benefits of being a cellulosic fibre, in that it is fully biodegradable and absorbent. It has high strength in both the wet and dry state. It blends well with fibres such as cotton, linen and wool. There are many published literature on production of lyocell fibres. In this section, the production, properties and application of lyocell fibres are discussed in detail.

3.2 Production Process

A diagram of the process flow in manufacturing of lyocellfibre is shown in Fig. 1.

The Lyocell process makes use of a direct solvent rather than indirect dissolution like the xanthation/regeneration route in the viscose process. Lyocell fiber is produced from dissolving pulp, which contains purified cellulose with little hemicellulose and no lignin. The rolls of pulp are broken into one-inch squares and dissolved in *N*-methylmorpholine-*N*-oxide (Fig. 2), giving a solution called "dope." The cellulose solution is then filtered and pumped through spinnerets. The fibres are drawn in air to align the cellulose molecules, giving the Lyocell fibres its characteristic high strength. The fibres are then immersed into a water bath that contains some dilute NMMO in a steady state concentration. The fibers are then washed with de-mineralized water. The Lyocell fibres are then taken for drying zone to evaporate the water from it.

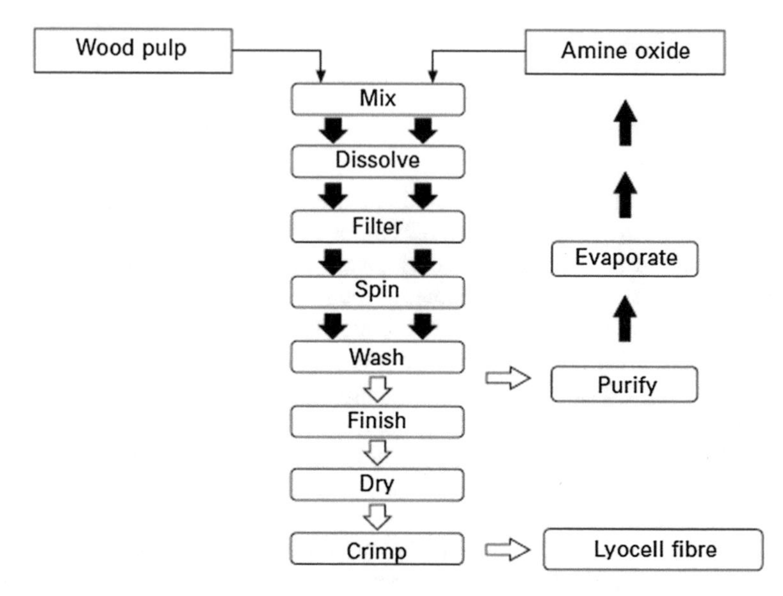

Fig. 1 Process flow in Lyocell fibre manufacturing. Reproduced from White et al. (2005)

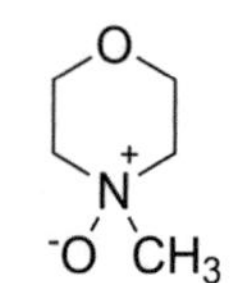

Fig. 2 Chemical structure of *N*-methylmorpholine-*N*-oxide

The production then follows the same route as with other kinds of fibers such as viscose. The fibre strands pass to a finishing area, where a lubricant, which may be soap or silicone or other agent is applied on the fibres depending on its future use. The dried, finished fibres are at this stage in a form called tow, a large untwisted bundle of continuous lengths of filament. The bundles of tow are then taken to a crimping machine that compresses the fibre, giving it texture and bulk. The NMMO used to dissolve the cellulose and set the fibre after spinning is recycled to the extent of 98%. Since the waste generated out of the entire process is very less, this process is relatively eco-friendly when compared to the conventional viscose process.

3.3 Lyocell Fibre Properties

Lyocell fibres possess many properties that are similar to other cellulosic fibres such as cotton, linen, ramie and viscose rayon. Lyocell shows a dry tenacity significantly higher than other cellulosics and approaching that of polyester. The lyocell fiber has a highly crystalline structure in which crystalline domains are continuously dispersed

Table 2 Properties of lyocell fibres vis-a-vis viscose, cotton and polyester (White et al. 2005; Chavan and Patra 2004)

Property	Tencel®	Viscose	Cotton	Polyester
Fineness (dtex)	1.7	1.7	–	1.7
Dry tenacity (cN/tex)	38–42	22–26	20–24	55–60
Dry elongation (%)	14–16	20–25	7–9	25–30
Wet tenacity (cN/tex)	34–38	10–15	26–30	54–58
Wet elongation (%)	16–18	25–30	12–14	25–30
Moisture regain (%)	11.5	13	8	0.4

along the fiber axis. This offers good wet strength and the fibre retains 85% of its dry strength in wet state is shown in Table 2. In addition, Lyocell has a high modulus that leads to low shrinkage in water. Further, it shrinks less when wetted by water and dried than other cellulose fibers such as cotton and viscose rayon (White et al. 2005; Chavan and Patra 2004).

3.4 *Textile Production*

The lyocell fibre has been produced either as a continuous tow or cut staple fibre and it is then converted to yarns and fabrics by a range of conventional textile processes. The most common way of using lyocell fibre is as cut staple, with 1.4 and 1.7 dtex fibres cut to 38 mm and converted into a spun yarn. The fibre can be processed on conventional machinery, with minimal changes in settings of machines for optimizing performance. The SEM picture of yarn cross section of a Tencel spun yarn is shown in Fig. 3. The spun yarn can be woven or knitted like any other fabric, and may be given a variety of finishes depending on the end use requirements.

3.5 *Wet Processing of Lyocell*

The processing of a greige lyocell fabric, like other cellulosic fabrics, also begins with the cleaning process. The impurities present in it are predominantly those introduced during fabric making. Hence, for woven fabrics, it is mainly size while for knits, yarn lubricants and knitting oil comprise the impurities. In 100% lyocell fabrics of woven qualities, PVA or a mixture of PVA and polyacrylates are used as sizing material.

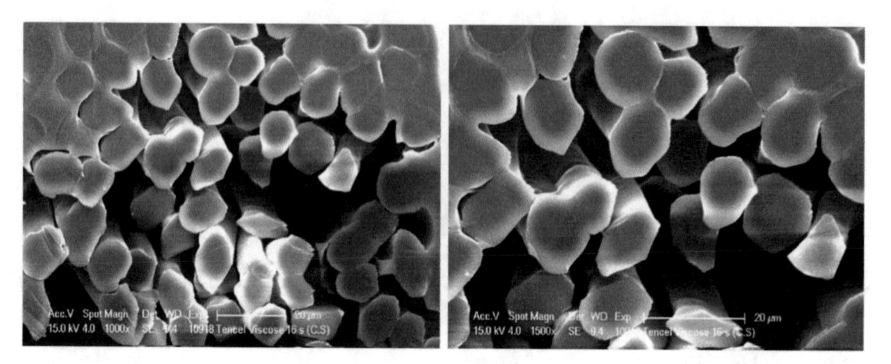

Fig. 3 SEM images of Tencel spun yarn cross section

Hence, the removal of these water-soluble added impurities does not require much of chemicals. But woven fabrics of lyocell/cotton blend are often sized with PVA, polyacrylates and starch. In such cases, desizing is done with enzymes to ensure removal of starch. However, lyocell/cotton knits don't require any desizing treatment for obvious reasons. Regarding the pre-treatments of scouring and bleaching, lyocell doesn't require any rigorous scouring. It can however be bleached, if required. A single-stage scouring and peroxide bleaching can be conveniently done. Lyocell/Tencel is a cellulosic fibre and therefore can be dyed with any class of dyestuff suitable for other cellulosics. In fact, it is observed that the dye yield on tencel is found to be greater than that on cotton, modal and viscose (Chavan and Patra 2004).

3.6 Fibrillation

Fibrillation of lyocell and other regenerated cellulosic fibers occurs during its wet processing and usage (Udomkichdecha et al. 2002). This leads to consumer fabric dissatisfaction phenomenon such as pilling. Fibrillation is the splitting of fibril bundles and its subsequent exposure to the fiber surface. When abraded in the wet state, these exposed fibrils may form aggregates (as pills) on the fiber surface or break away as lint. This also causes problems during dyeing and finishing processes. Lyocell fibers possess 90% or higher crystallinity, and large crystalline regions with the longitudinal orientation of the crystallites and low lateral cohesion between the fibrils (Mortimer and Peguy 1996; Valldeperas et al. 2000). In the wet state, the water penetrates inside the fibrillar bundles, resulting fiber swelling and breakage of hydrogen bonds. After losing hydrogen bonds, the bundled fibrils try to relax. This leads to fibril splitting and subsequent exposure of the fibrils onto the fiber surface. Fibrillation of the fiber surface leads to the formation of "pills" and "peach skin" effect in clothing.

Some of the ways to reduce fibrillation in lyocell fibres are; (1) Controlling spinning parameters (2) Controlling wet process parameters (3) Use of crosslinking agents (4) Enzymatic treatments and (5) Dyeing with polyfunctional reactive dyes and binders.

Lenzing has introduced Tencel A100 and Tencel LF as a non-fibrillating variants of Tencel which was developed to meet identified market needs for a Tencel fibre that did not fibrillate and therefore would be easier to dye and finish, which could be processed on conventional processing equipment and would produce a different aesthetic in fabrics. Fibrillation is prevented by cross-linking the fibre during manufacture before it is dried. The result is a fibre that processes well. The cross-linker used is TAHT (triacrylamidotrihydrotriazine). This is a trifunctional molecule that reacts readily with hydroxyl groups in alkaline conditions at elevated temperature. The molecule has chemistry very similar to reactive dyes (Burrow 2005). It can penetrate the amorphous regions of the water swollen cellulose where it reacts with one or more cellulose groups. When it reacts with hydroxyl groups on two cellulose chains it binds them together and prevents them from splitting apart.

3.7 Applications of Lyocell

Lyocell is more expensive fibre to produce than cotton or viscose rayon. The fibre is much commercialized and is being used in many apparel and home textile applications. Staple fibres are used in clothes such as denim, underwear, casual wear, bed sheets and towels. The filament fibres are used in women's clothing and men's dress shirts so as to get a silkier appearance. Lyocell can be blended with a variety of other fibres such as silk, cotton, rayon, polyester, linen, nylon, and wool. Both binary and tertiary blended materials are presently in use. Some examples are; 65/35 cotton/lyocell, 50/50 cotton/lyocell, 49/46/5 cotton/lyocell/wool etc. (Kilic and Okur 2011). Lyocell is also used in technical applications like conveyor belts, specialty papers, and medical dressings.

3.8 Eco-friendliness of Fibre

In addition to their aesthetic and performance benefits, lyocell fibres also offer the potential for very attractive environmental characteristics. One of the key development targets for lyocell fibre was to deliver a product offering significant benefits in terms of low environmental impact and sustainability. Lyocell is considered to be a 'sustainable fibre' because: (1) The forests that provide the raw material for lyocell are always being replenished. (2) Other materials used in the fibre production process are re-cycled with very little loss. (3) The fibre is biodegradable. Lyocell has distinct advantage over viscose rayon in the sense that rayon manufacturing generates lot of air and water pollution, uses catalytic agents containing cobalt or manganese, and creates a strong, unpleasant odor. On the other hand, the chemicals that are used to manufacture lyocell fibers are nontoxic. In addition, the cellulose pulp used for lyocell is treated in a closed loop process in which these solvents are recycled with a recovery rate of more than 98%. Undoubtedly lyocell fibres are better compared to the standard viscose rayon from environmental perspective.

3.9 Recent Innovations

The environmentally friendly production process sets lyocell apart from other man-made cellulosic fibre types, and this will gain in importance over the coming years. Lyocell fiber have some key characteristic over other cellulosic fibers; such as high dry and wet tenacity and high wet modulus. The introduction of the new non-fibrillating fibre types of Tencel® A100 and Lenzing lyocell LF have seen impressive growth since their introduction. The main focus in future will be to reduce the cost of production of this fibre and expanding its markets into new product segments. Lenzing has recently launched a Tencel ecological fibre produced from cotton fabric wastes. According to the manufacturer, the latest next-generation Tencel fibre combines the best of two worlds, recycling cotton waste fabrics and using the most sustainable Tencel technology to create an ecological wood-based fibre.

4 Regenerated Wool Keratin Fibres

4.1 Introduction

Currently, protein-based biomaterials are widely used in tissue engineering, regenerative medicine and various biomedical applications (Altman et al. 2003; Goo et al. 2003). Wool processing and poultry industries are generating significant amount of waste material, which contain sustainable material named keratin-protein. Keratin derived from wool, feathers and horns has received good attention for application in biomedical fields due to their biocompatibility, cellviability and biodegradability (Yamouchi et al. 1998). Wool processing industry and chicken poultry industries are producing more than three million tonnes of keratin-based bio-wastes (Moncrieff 1975). Purified and regenerated keratin from wool waste is widely used for biomedical applications (Maclaren and Milligan 1981; Tonin et al. 2006). Aluigi et al. (2007) stated that keratin needs cross-linking agents to improve mechanical strength and processing feasibilities. Alemdar et al. (2005), Liu et al. (2004) used formic acid as a solvent for the preparation of keratin/fibroin, keratin/nylon 6 and nylon 66 polymer blend and structural studies (Aluigi et al. 2007) revealed the thermal stability of keratin and stabilization of β-sheet structure. Currently eco-friendly fibres are slowly replacing the synthetic fibres usage due to environmental concerns and driving force to do research in those areas (Poole et al. 2009). Various survey results revealed that environment friendly fibres/products must meet the functional requirements and performances similar to nonecofriendly products (Thiry 2007; Anon 2007b) and increasing sale argument (Anon 2007a). Reddy and Yang (2007) stated that ligno-cellulosic by products (corn stalks, husks, leaves) from agriculture may replace the usage of cellulosic fibres in future. Wool and silk are protein fibres having good mechanical strength at dry condition due to inter-chain bonding (Hearle 2007).

4.2 *Properties of Keratin*

McKittrick et al. (2012) explained the stability of keratin protein to sunlight, biological attack, mechanical distortion and water. The derived keratin from wool, chicken feather, and horns are used for fabrication of films, scaffold and thermoplastic sheets for biomedical and industrial applications (Dyer and Ghosh 2013; Hill et al. 2010; Aluigi et al. 2007; Li et al. 2012). A reduction in thermal stability, absence of α-helix structure (Li and Wang 2013) and formation of disordered structures (Xie et al. 2005; Idris et al. 2014) were found in thermally heated wool fibre. Disappearance of bands between 37 and 75 KDa, appearance of new bands in the range of 20–30 KDa and degradation of low molecular mass protein was observed at 180 °C in a super-heated steamer (Ghosh et al. 2014; Bertini et al. 2013). Ghosh et al. (2014) studied the structure and properties of wool keratin while using ionic liquid as solvent and results revealed improved thermal processing properties and this could be a route for production of bio-resin for industrial applications. Rouse and Dyke (2010), Aluigi et al. (2008) and many scientists used wool keratin material for drug delivery, wound healing, tissue engineering and various biomedical applications.

Wei et al. (2013) studied the effect of pH and solvent assisting agent on dissolution percentage of wool fibre in reductive method and results revealed that high pH value (>12) and high concentration of mental salts degrades the wool fibre. The ionic liquid 1-allyl-3-methylimidazolium chloride $[AMM]^+ Cl^-$ possesses positive dissolution property for wool keratin fibres and structural studies of regenerated keratin film shows broken disulfide bonds, absence of α-helix structure and decrease in thermal stability property (Li et al. 2012). Recycling and effective utilization of short and crude fibres from weaving industry is utmost important else wool waste from textile industry pollute the environment (Lv and Yu 2010). Wool keratin protein has 3D structure due to highly crosslinked bonds such as hydrogen, salt, disulphide and other bonds (Hames and Hooper 2003). Protein fibre properties were affected by cross-linking of inter-chain bonds (Ziabicki 1967). Chicken feather keratins have β-sheet conformation and rich in cystein and hydrophobic residues and widely used as films (Yin et al. 2007; Tanabe et al. 2004), coating material (Schrooyen et al. 2001), and composite materials (Barone and Schmidt 2005) due to their inherent properties of biocompatibility, biodegradability, nontoxicity and wound healing nature (Yin 2007; Halford 2004; Schrooyen et al. 2001). Keratin derived from chicken feather and wool fibres have high molecular mass or molecular cut-off (Gillespie 1990), high cysteine content (Bradbury et al. 1967), and helical configuration (Plowman 2003).

4.3 Recent Innovations

Zeng and Qi (2011) stated that dissolution supporting agents such as lithium chloride and calcium chloride increased the recovery rate or yield percentage of keratin from wool fibre. Currently, ionic liquids are used as green solvent for keratin biomaterial due to its unique properties (thermal stability, chemical resistance), apart from being used for polymer modifications and polymerization process (Winterton 2006; Kubisa 2005; Sun et al. 2009; Xie et al. 2005; Phillips et al. 2004). The blends of wool protein and polyacrylonitrile fibre shows improved moisture regain property and good biocompatibility with human skin. Keratin polymer may be used for the development of innovative biomaterials such as biofilms, biocomposites and gels for biomedical applications such as drug delivery systems and scaffolds for tissue engineering (Silva et al. 2014).

5 Chitosan Polymer/Fibre

5.1 Introduction

Biodegradable polymers are gaining much importance in the current era and they are derived from various sources. For the past 5 years, natural wastes like cell walls of algae, outer skeleton of insects, shells of crab, chicken feather, human hair and much more are used as sources/raw materials for the preparation of biodegradable polymers. Among the biodegradable biopolymers, chitosan is widely used biopolymer in biomedical and pharmaceutical fields due to its unique properties such as biodegradable, non-toxic, non-allergenic bi-functional and biocompatible in nature (Anitha et al. 2014; Ahmed and Ikram 2016; Shanmugasundaram 2012; Abd Elgadir et al. 2015). Chitin and chitosan are natural biopolymers derived from waste products of the shrimp canning and crabbing industry (Dutta et al. 2004). Chitosan is the deacetylated form of chitin, composed of β-(1-4)-2 acetamido-2-deoxy-D-glucose and β-(1-4)-2-amio-2-deoxy-D-glucose depicted in Fig. 1 (Dutta et al. 2004). Chitin is a natural polysaccharide material abundantly available on earth next to cellulose (Prashanth and Tharanathan 2007). Chitosan was first discovered by Rouget in 1859, when he boiled chitin solution in a potassium hydroxide solution (Muzzarelli 1977). Ribgy (1934) obtained two patents in 1934, one for producing chitosan from chitin and other for making fibres and films from chitosan. Clark and smith (1936) published first X-ray pattern of well-oriented chitosan fibres. Salmon and Hudson (1997) reported that chitin possesses acetamide group in C2 position and purity, morphology and molecular weight are dependent on its nature.

5.2 Properties of Chitosan

Chitosan is a polycationic polymer and by chemical modification (Li et al. 1992) they can be easily converted into different forms such as films, fibres (nano and micro), gels, membranes, etc. Chitosan scaffold shows poor mechanical strength and unstable besides the advantages of porosity, flexibility, ease of processing, and biocompatibility (Martino et al. 2005). However, mechanical strength of scaffold can be improved by blending alginate polymer (Li et al. 2005), poly(lactic-co-glycolic) acid (Jiang et al. 2006). Chitosan is soluble in acidic solvents (Leedy et al. 2011) and it contains minimum number of *N*-acetyl-2-amino-2-deoxy-D-glucose group (Khor and Lim 2003). A rigid and unbranched structure found in chitin and chitosan polymer is due to presence of β-1,4-linkage (Anitha et al. 2014). Muzzarelli and Muzzarelli (2005) reported that chitosan biopolymer possesses free amino groups which exhibits bacterio-static and fungistatic effects. Practical applications of chitin are limited due to its poor solubility. Chitosan is used to remove heavy metals present in waste water/effluents (Rinaudo 2006). Chitosan is ability to form ionic complexes with anionic natural and synthetic species like proteins, DNA and negatively charged poly (acrylic acid) synthetic polymers (Pavinatto et al. 2010; Takahashi et al. 1990; Kim et al. 2007; Croisier and Jerome 2013). These polysaccharide polymers have the ability to form stable covalent bond with other species due to presence of amino and alcohol groups (Croisier and Jerome 2013).

5.3 Applications of Chitosan Polymer

5.3.1 Wound Healing (Films and Scaffold)

Non-woven mats, sponges and film form of chitosan exhibit an increase in wound healing efficiency. Chitosan polymer is being used in various forms such as cotton fibre type (Ueno et al. 1999), Chitosan-alginate PEC membrane (Wang et al 2002), chitosan/collagen scaffold (Ma et al. 2003), collagen-chitosan membrane (Guo et al. 2011), chitosan coated cotton gauze (Shanmugasundaram and Gowda 2012b), polylactic acid bandage (Shanmugasundaram and Gowda 2012a) and bamboo bandage (Shanmugasundaram and Gowda 2011), chitosan-nanofibrin bandage (Sudheesh Kumar et al. 2013), collagen-chitosan matrix gel (Judith et al. 2012) and chitosan-gelatin scaffolds to heal normal wound, burn wounds and excisional wounds. Electrospun fibrous mat was found to promote cell attachment and proliferation in wound healing process (Zhou et al. 2008). Chitosan in the form of hydrogel can act as positive antibacterial material for wound healing applications (Fujita et al. 2004). A research group developed easily stripped-off chitosan bandage for wound healing applications (Chen et al. 2005a, b). Silver sulfadiazine release chitosan membrane for wound healing (Fwu et al. 2003). Chitosan immobilized with drug to obtain wound healing effect (Joshua et al. 2008). Hani and Satya (2013) reported that wound healing is a complex process. Chitosan/heparin complex substance shows effective wound healing ability (Kweon et al. 2003).

5.3.2 Drug Delivery/Releasing Systems

Chitin, chitosan and its blends were used for sustained drug delivery system (Shilpa et al. 2003; Felt et al. 1998), controlled drug releasing systems (Surini et al. 2003; Bernardo et al. 2003). Thacharodi and Rao developed chitosan membrane for nifedipine (Thacharodi and Rao 1993, 1996) and propranolol hydrochloride (Thacharodi and Rao 1995) drug delivery system. Chitosan films have been developed for potential delivery of antibiotics drug for wound healing (Noel et al. 2008). Lopez et al. (1998) introduced chitosan/ethylcellulose bi-layered device for delivery of buccal drug. Chitosan/hydroxyapatite scaffolds were found to be efficient in release of dexamethasone drug to cure allergic disorders (Tigli et al. 2009) and controlled release of growth factor to promote cell growth during wound healing process (Ho et al. 2009).

5.3.3 Tissue Engineering and Regenerative Medicine

Zhang and Zhang (2001) developed microporous chitosan/calcium phosphate composite scaffold for tissue engineering. Many scientists have developed porous scaffold (Jarry et al. 2001; Madihally and Matthew 1999a, b) composite scaffold (Kast et al. 2003) and artificial skin (Mucha 1997) for bone tissue engineering. Toskas et al. (2013) produced pure chitosan microfibers/fabrics for regenerative medicine applications. Chitosan scaffolds were produced in combination with polycaprolactone (Wu et al. 2010), collagen (Shi et al. 2009), poly(butylenes succinate) (Shi et al. 2009), Hydroxyapatite (Anitha et al 2014), silkfibroin (Wang and Li 2007), carbon nanotube (Venkatesan et al. 2011) and hydroxyapatite/CMC composite to regenerate bone tissue (Jiang et al. 2008) and various biomedical applications. Many researchers published a detailed review report on chitosan based scaffolds for wound healing applications (Ahmed and Ikram 2016; Croisier and Jerome 2013; Rinaudo 2006; Anitha et al. 2014).

5.3.4 Water Treatment

Muzzarelli (1973) reported the effectiveness and ability of chitosan polymer to chelate harmful metal ions from wastewater and considered as safe polymer for human use (Hirano et al. 1989). Chitosan used as sorbents for removal of metal ions in acidic medium (Weltrowski et al. 1996), absorbents for removal of colour from effluents (Bhavani and Dutta 1999) and wastewater treatment (Sridhari and Dutta 2000). It can also be used to remove petroleum from wastewater, arsenic from drinking water and purification of potable water (Dutta et al. 2004). Hirano et al.

(1989) recovered uranium from sea, river and lake water and No and Meyers (1989) recovered amino acids from wastewater. Crini (2005) used chitosan as adsorbents in wastewater treatment.

5.3.5 Other Industries

Wu et al. (1978) reported that chitosan has the ability to remove protein from cheese whey. An increase in paper bursting strength and folding endurance can be obtained by treating paper with 1% chitosan solution (Muzzarelli 1983). A layer of chitosan was coated on photographic paper to increase antistatic properties (Aizawa and Noda 1988). Tokura et al. (1988) produced antithrombogenic medical products from chitosan and its derivatives. Some researchers have developed anticoagulant membrane using chitosan (Chandy and Sharma 1989). The cholesterol in blood can be decreased by taking chitosan (power, solution) as medicine (Muzzareli 1985).

5.4 Recent Innovations

Freudenberg Performance Materials is one of the global manufacturers of innovative technical textiles such as apparel, automotive, building materials, hygiene, medical, etc. They developed innovative products such as antimicrobial foam or nonwovens made from chitosan fibers to heal wounds like pressure sores and diabetic foot. Trusetal Verbandstoffwerk GmbH developed chitoderm® plus (superabsorber and the bacteriostatic chitosan coating) to treat lightly to heavily exudating wounds, acute and chronic wounds. Axiostat® is a pure chitosan dressing made chitosan for medical application. Axio uses innovative technology to purify chitosan so that a high performance dressing material is made available to heal chronic wounds.

6 Conclusions

Sustainability is a key driver for future fibre market. The environmental friendliness of fibres during its production, processing and disposal is going to decide the commercial acceptance of these materials. The lyocell process for regenerated cellulose fibre was a huge success and the lyocell fibre with a brand name of Tencel is already enjoying a good market share. Many regenerated rayon manufacturers have switched over to more eco-friendly processes for production. Polylactic acid for plastic and composite applications is on the rise. However, the textile applications are yet to pick up for this interesting fibre. The fibre manufacturing process needs further improvement and the cost of manufacturing needs to be brought down. Presently, only few manufacturers across the world produce PLA fibres. This fibre has a tremendous potential to replace the PET polyester in future. The regenerated wool keratin fibre and chitosan polymer/fibres have been researched by many and have shown good potential for specific applications. However, the commercial acceptance of these materials is dependent on their availability in huge quantities, their suitability for some unique end-use purposes and on cost aspects.

References

Abd Elgadir M, Salim Md, Uddin Sahena Ferdosh et al (2015) Impact of chitosan composites and chitosan nanoparticle composites on various drug delivery systems: a review. J Food Drug Anal 23:619–629

Ahmed S, Ikram S (2016) Chitosan based scaffolds and their applications in wound healing. Achievements Life Sci 10:27–37

Aizawa Y, Noda T (1988) Antistatic photographic paper. Jpn KokaiTokkyoKoho JP 63:189

Alemdar A, Iridag Y, Kazanci M (2005) Int J Biol Macromol 35:151–153

Altman GH, Diaz F, Jakuba C, Calabro T, Horan RL, Chen J et al (2003) Biomaterials 24:401

Aluigi A, Zoccola M, Vineis C, Tonin C, Ferrero F (2007) Study on the structure and properties of wool keratin regenerated from formic acid. Int J Biol Macromol 41:266–273

Aluigi A, Vineis C, Ceria A, Tonin C (2008) Composite biomaterials from fibre wastes: characterization of wool-cellulose acetate blends. Compos A 39:126–132

Anitha A, Sowmya S, Sudheesh Kumar PT, Deepthi S et al (2014) Chitin and chitosan in selected biomedical applications. Prog Polym Sci 39:1644–1667

Anon (2007a) Green textiles in demand WSA performance and sports materials. World Trades Publishing, Liverpool, pp 35–37

Anon (2007b) How green is becoming the new black in the textile chains, WSA performance and sports materials. World trade publishing, Liver Pool, pp 35–37

Baig GA, Carr CM (2014) Polish J Chem Tech 16(3):45–50

Barone J, Schmidt W (2005) Comps Sci Technol 65:173–181

Bernardo MV, Blanco MD, Sastre RL, Teijon C(2003) Sustained release of bupivacaine from devices based on chitosan. II Farmaco 58:1187

Bertini F, Canetti M, Patrucco A et al (2013) Wool keratin-polypropylene composites: properties and thermal degradation. Polym Degrad Stab 98:980–987

Bhavani KD, Dutta PK (1999) Physico-chemical adsorption properties on chitosan for dyehouse effluent. Am Dyestuff Reporter 88:53

Bradbury JH, Chapman GV, King NLR, (1967) Chemical composition of the histological components of wool. In Crewther WG (ed) Symposium on fibrous proteins Australia. Butterworths, Sydney pp 368–372

Burrow TR (2005) Recent advances in chemically treated lyocellfibers. LenzingerBerichte 84:110–115

Chandy T, Sharma CP (1989) J Colloid Interface Sci 130:331–340

Chavan RB, Patra AK (2004) Development and processing of lyocell. Indian J Fibre Text Res 29:483–492

Chen KS, Ku YA, Lee CH, Lin HR (2005a) Immobilization of chitosan gel with cross-linking reagent on PNIPAAm/gel/PP non-woven composite surface. Mater Sci Eng 25:472–478

Chen SP, Wu GZ, Zeng HY (2005b) Preparation of high antimicrobial activity thiourea chitosan-Ag+ complex. Carbohyd Polym 60:33–38

Clark GL, Smith AF (1936) X-ray diffraction studies of chitin, chitosan and derivatives. J Phys Chem 40:863–879

Crini G (2005) Recent developments in polysaccharide-based materials used as adsorbents in wastewater treatment. Prog Polym Sci 30:38–70

Croisier F, Jerome C (2013) Chitosan-based biomaterials for tissue engineering. Eur Polym J 49:780–792
Drumright RE, Gruber PR, Henton DE (2000) Adv Mater 12(23):1841
Dutta PK, Dutta J, Tripathi VS (2004) Chitin and chitosan: chemistry, properties and applications. J Sci Ind Res 63:20–31
Dyer JM, Ghosh A, (2013) Keratin nanomaterials: development and applications. In: Aliofkhazraei M (ed) Handbook of functional nanomaterials. Properties and commercialization, vol 4. Nova Publishers, New York
Farrington DW, Lunt J, Davies S et al (2005) Poly (lactic acid) fibers. In: Blackburn RS (ed) Biodegradable and sustainable fibers. Woodhead publishing, Cambridge, pp 191–220
Felt O, Buri P, Gurny R (1998) Chitosan: a unique polysaccharide for drug delivery. Drug Deliv Ind Pharm 24:979
Fujita M, Kinoshita M, Ishihara M, Kanatani Y et al (2004) Inhibition of vascular prosthetic graft infection using a photocrosslinkable chitosan hydrogel. J Surg Res 121:135–140
Fwu LM, Yu BW, Shin SS, Chao AC (2003) Asymmetric chitosan membranes prepared by dry/wet phase separation: a new type of wound dressing for controlled antibacterial release. J Membr Sci 212:237–254
Ghosh A, Clerens S et al (2014) Thermal effects of ionic liquid dissolution on the structures and properties of regenerated wool keratin. Polym Degrad Stab 108:108–115
Gillespie JM (1990) The protein of hair and other hard α-keratin. In Goldman RD, Steinert PM (eds) Cellular and molecular biology of intermediate filaments. Plenum Press, New York, pp 95–128
Goo HC, Hwang YS, Choi YR, Suh H (2003) Biomaterials 24:5099
Guo R, Xu S, Ma I, Huang A, Gao C (2011) The healing of full-thickness burns treated by using plasmid DNA encoding VEGF-165 activated collagen-chitosan dermal equivalents. Biomaterials 32:1019–1031
Guruprasad R, Vivekanandan MV, Arputharaj A, Saxena S, Chattopadhyay SK (2015) Development of cottonrich/polylactic acid fiber blend knitted fabrics for sports textiles. J Ind Text 45(3):405–415
Halford B (2004) Going beyond feather dusters. Chem Eng News 36–39
Hames BD, Hooper NM (2003) Biochemistry, 2nd edn. Science Press, Beijing
Hani S, Satya P (2013) Complements and the wound healing cascade: an updated review. Plast Surg Int 1–7
Hearle J (2007) J Mater Sci 42:8010–8019
Hill P, Brantley H, Van Dyke M (2010) Some properties of keratin biomaterials: kerateines. Biomaterials 31:585–593
Hirano S, Seino H, Akiyama Y, Nonaka I (1989) Polym Mater Sci Eng 59:897–901
Ho YC, Mi FL, Sung HW, Kuo PL (2009) Heparin-functionalized chitosan-alginate scaffolds for controlled release of growth factor. Int J Pharm 376:69–75
Idris A, Vijayaraghavan UA, Rana AF et al (2014) Dissolution and regeneration of wool keratin in ionic liquids. Green Chem 16:2857–2864
Jarry C, Chaput C, Chenite A et al (2001) Effects of stem sterilization on thermogelling chitosan-based gels. J Biomed Mater Res 58:127
Jiang T, Abdel-Fattah WI, Laurencin CT et al (2006) In vitro evaluation of chitosan/poly(lactic acid-glycolic acid) sintered microsphere scaffolds for bone tissue engineering. Biomaterials 27:4894–4903
Jiang L, Wang X, Li Y et al (2008) Preparation and properties of nano-hydroxyapatite/chitosan/carboxymethyl cellulose composite scaffold. Carbohydr Polym 74:680–684
Joshua SB, Kerr HM, Howard NES, Eccleston M (2008) Wound healing dressings and drug delivery systems: a review. J Pharm Sci 97:2892–2923
Judith R, Nithya M, Rose C, Mandal AB (2012) Biopolymer gel matrix as acellular scaffold for enhanced dermal tissue regeneration. Biologicals 40:231–239
Kast CE, Frick W, Losert U, Schnurch AB (2003) Chitosanthioglycolic acid conjugate: a new scaffold material for tissue engineering. Int J Pharm 256:183

Khor E, Lim LY (2003) Implantable applications of chitin and chitosan. Biomaterials 24:2339–2349
Kilic M, Okur A (2011) The properties of cotton-tencel and cotton-promodal blended yarns spun in different spinning systems. Text Res J 81(2):156–172
Kim T-H, Jiang H-L, Jere D et al (2007) Chemical modification of chitosan as a gene carrier in vitro and in vivo. Prog Polym Sci 32:726–733
Kubisa P (2005) Ionic liquids in the synthesis and modification of polymers. J Polym Sci Part Polym Chem 43:4675–4683
Kweon DK, Song SB, Park YY (2003) Preparation of water-soluble chitosan/heparin complex and its applications as wound healing accelerator. Biomaterials 24:1595–1601
Leedy M, Martin H, Norowski P et al (2011) Use of chitosan as a bioactive implants coating for bone-implant applications. In: Jayakumar R, Prabaharan M, Muzzarelli RAA (eds) Chitosan for biomaterials II. Advances in polymer science. Springer, Berlin, pp 129–65
Li R, Wang D (2012) Preparation of regenerated wool keratin films from wool keratin-ionic liquid solutions. J Appl Polym Sci 127:2648–2652
Li R, Wang D (2013) Preparation of regenerated wool keratin films from wool keratin-ionic liquid solutions. J Appl Polym Sci 127:2648–2653
Li Q, Dunn ET, Grandmaison EW, Goosen MFA (1992) Applications and properties of chitosan. J Bioact Compatible Polym 7:370
Li Z, Ramay HR, Hauch KD et al (2005) Chitosan-alginate hybrid scaffolds for bone tissue engineering. Biomaterials 26:3919–3928
Li Q, Zhu L, Liu R, Huang D, Jin X et al (2012) Biological stimuli responsive drug carriers based on keratin for triggerable drug delivery. J Mater Chem 22:19964–19973
Liu Y, Shao Z, Zhou P, Chen X (2004) Polymer 45:7705–7710
Lopez CR, Portero A, Vila-Jato JL, Alonso MJ (1998) Design and evaluation of chitosan/ethylcellulose mucoadhesive bilayered devices for buccal drug delivery. J Controlled Release 55:143–152
Lunt J, Bone J (2001) AATCC Rev 1(9):20
Lv LH, Yu YL (2010) J Zhou Tekstil 29:201
Ma I, Gao C, Mao Z, Zhou J, Shen J et al (2003) Collagen/chitosan porous scaffold with improved biostability for skin tissue engineering. Biomaterials 24:4833–4841
Maclaren JA, Milligan B (1981) Wool science, vol 1. Science Press, pp 1–18
Madihally SV, Matthew HWT (1999a) Porous chitosan scaffolds for tissue engineering. Biomaterials 20:1133–1142
Madihally SV, Matthew HWT (1999b) Porous chitosan scaffolds for tissue engineering. Biomaterials 20:1133
Martino AD, Sittinger M, Risbud MV (2005) Chitosan: a versatile biopolymer for orthopaedic tissue-engineering. Biomaterials 26:5983–5990
McKittrick J, Chen PY, Bodde SG, Yang W (2012) The structure, functions and mechanical properties of keratin. JOM 64:449–468
Moncrieff RW (1975) Man made fibres, vol 11, 6th edn. Butterworths Scientific, London, p 231
Mortimer SA, Peguy AA (1996) Methods for reducing the tendency of lyocell fibers to fibrillate. J ApplPolymSci 60(3):305–316
Mucha M (1997) Rheological characteristics of semi-dilute chitosan solutions. Macromol Chem Phys 198:471
Muzzarelli RAA (1973) Natural Chelating Polymers. Pergamon of Canada Ltd., Toronto, pp 83–95
Muzzarelli RAA (1977) Chitin. Pergamon Press, New York
Muzzarelli RAA (1983) Carbohyd Polym 3:53–75
Muzzareli RAA (1985) In: Aspinall GO (ed) The polysaccharides, vol 3. Academic Press Inc., London, pp 417–451
Muzzarelli RAA, Muzzarelli C (2005) Chitosan chemistry: relevance to the biomedical sciences. Adv Polym Sci 186:151–209
Nakumara T (2003) Int Text Bull 4:68
No HK, Meyers SP (1989) J Agric Food Chem 54:60–62

Noel SP, Courtney HS, Burngaardner JD, Haggard WO (2008) Chitosan films a potential local drug delivery system for antibiotics. Clin Orthop Relat Res 466:1377–1382
Parmar MS, Singh M, Tiwari RK, Saran S (2014) Study on Flame retardant properties of poly(lactic acid) fibre fabrics. Indian J Fibre Text Res 39:268–273
Pavinatto FJ, Oliveira CL (2010) Chitosan in nanostructured thin films. Biomacromolecules 11:1897–1908
Phatthalung IN, Sae-be P, Suesat J, Suwanruji P, Soonsinpai N (2012) Investigation of the optimum pretreatment conditions for the knitted fabric derived from PLA/cotton blend. Int J Biosci Biochem Bioinf 2(3):179–182
Phillips DM, Drummy LF, Conrady DG et al (2004) Dissolution and regeneration of bombyx mori silk fibroin using ionic liquids. J Am Soc 126:14350
Plowman JE (2003) J Chromatogr B Anal Technol Biomed Life Sci 787:63–76
Poole AJ, Church JS, Huson MG (2009) Environmentally sustainable fibers from regenerated protein. Biomacromolecules 10:1–7
Prashanth KVH, Tharanathan RN (2007) Chitin/chitosan: modifications and their unlimited application potential—an overview. Trends Food Sci Technol 18:117–131
Reddy N, Yang Y (2007) J Polym Environ 15:81–87
Ribgy GW (1934) Substantially undergraded deacetylated chitin and process for producing the same. U.S. Patent 2,040,880
Rinaudo M (2006) Chitin and chitosan: properties and applications. Prog Polym Sci 31:603–632
Rouse JG, Dyke MV (2010) Materials 3:999
Salmon S, Hudson SM (1997) Crystal morphology biosynthesis and physical assembly of cellulose, chitin and chitosan, review. J Macromol Sci Macromol Chem Phys C 37:199
Schrooyen P, Dijkstra P, Oberthur P et al (2001) Agric. Food Chem 49:221–230
Shanmugasundaram OL (2012) Development and characterization of cotton and organic cotton gauze fabric coated with biopolymers and drugs for wound healing. Indian J Fiber Text Res 37:146–150
Shanmugasundaram OL, Gowda RVM (2011) Development and characterization of bamboo gauze fabric coated with polymer and drug for wound healing. Fibers Polym 12:15–20
Shanmugasundaram OL, Gowda RVM (2012a) Development and characterization of polylactic acid bandage coated with biopolymers and drugs for wound healing. J Text Inst 103:508–516
Shanmugasundaram OL, Gowda RVM (2012b) Development and characterization of cotton, organic cotton flat knit fabrics coated with chitosan, sodium alginate, calcium alginate polymers, and antibiotic drugs for wound healing. J Ind Text 42:156–175
Shi S, ChengX WangJ et al (2009) RhBMP-2 microspheres-loaded chitosan/collagen scaffold enhanced osseointegration: an experiment in dog. J Biomater Appl 23:331–346
Shilpa A, Agrawal SS, Ray AR (2003) Controlled delivery of drug from alginate matrix. J Macromol Sci Part C 43:187
Silva NHCS, Vilela C et al (2014) Protein-based materials: from sources to innovative sustainable materials for biomedical applications. J Mater Chem B 24:3707–3898
Sridhari TR, Dutta PK (2000) Synthesis and characterization of maleilated chitosan for dye house effluent. Indian J Chem Technol 7:198
Sudheesh Kumar PT, Raj NM, Praveen G et al (2013) In vitro and in vivo evaluation of microporous chitosan hydrogel/nanofibrin composite bandage for skin tissue regeneration. Tissue Eng 19:380–392
Sun P, Liu Z-T, Liu Z-W (2009) Particles from bird feather: a novel application of an ionic liquid and waste resource. J Hazard Mater 170:786–790
Surini S, Akiyama H, Morishita M, Nagai T, Takayama K (2003) Release phenomena of insulin from an implantable device composed of a polyion complex of chitosan and sodium hyaluronate. J Control Release 90:291
Takahashi T, Takayama K, Machida Y (1990) Characterization of polyion complexes of chitosan with sodium alginate and sodium polyacrylate. Int J Pharm 61:35–41
Tanabe T, Okitsu N et al (2004) Mater Sci Eng C 24:441–446

Thacharodi D, Rao KP (1993) Release of nifedipine through crosslinked chitosan membranes. Int J Pharm 96:33–39

Thacharodi D, Rao KP (1995) Collagen-chitosan composite membranes for controlled release of propranolol hydrochloride. Int J Pharm 120:115–118

Thacharodi D, Rao KP (1996) Rate-controlling biopolymer membranes as transdermal delivery systems for nifedipine: development and in vitro evaluations. Biomaterials 17:1307–1311

Thiry M (2007) If the environment is important, AATCC review american association of textile chemists and colourists. Research Triangle Park, North Carolina, pp 21–28

Tigli RS, Akman AC, Gumusderelioglu M, Nohutcu RM (2009) In vitro release of dexamethasone or bFGF from chitosan/hydroxyapatite scaffolds. J Biomater Sci Polym Ed 20:1899–1914

Tokura S, Itoyama M, Hiroshi S (1988) Partially sulphated chitosan oligomers immobilized on chitosan for antithrombogenic medical goods. Jpn KokaiTokkyoKoho JP, 63:89, 167

Tonin C, Zocola M, Aluigi A, Varesano A, Montarsolo A, Vineis C (2006) Biomacromol 7:3499–3540

Toskas G, Brunler R, Hund H et al (2013) Pure chitosan microfibers for biomedical applications. AUTEX Res J 13:134–138

Udomkichdecha W, Chiarakorn S, Potiyaraj P (2002) Relationships between fibrillation behavior of lyocell fibers and their physical properties. Text Res J 72(11):939–943

Ueno H, Yamada H, Tanaka I, Kaba N et al (1999) Accelerating effects of chitosan for healing at early phase of experimental open wound in dogs. Biomaterials 20:1407–1414

Valldeperas J et al (2000) Kinetics of enzymatic hydrolysis of lyocell fibers. Text Res J 70(11):981–984

Venkatesan J, Qian ZJ, Ryu B et al (2011) Preparation and characterization of carbon nanotube-grafted-chitosan-natural hydroxyapatite composite for bone tissue engineering. Carbohydr Polym 83:569–577

Wang L, Li C (2007) Preparation and physicochemical properties of a novel hydroxyapatite/chitosan-silk fibroin composite. Carbohydr Polym 68:740–745

Wang LS, Khor E, Wee Lim LY (2002) Chitosan-alginate PEC membrane as a wound dressing: assessment of incisional wound healing. J Biomed Mater Res 63:610

Wei Ju, Hao Lu, Sijin Xu et al (2013) Preparation and characterization of regenerated wool protein PAN blended fibers. Adv Mater Res 690–693:1461–1464

Weltrowski M, Martel B, Morcellet M (1996) Chitosan N-benzyl sulfonate derivatives as sorbents for removal of metal ions in an acidic medium. J Appl Polym Sci 59:647

White P, Hayhurst M, Taylor J, Slater A (2005) LyocellFibres. In: Blackburn RS (ed) Biodegradable and sustainable fibres. Woodhead Publishing, pp 157–190

Winterton N (2006) Solubilisation of polymers by ionic liquids. J Mater Chem 16:4281–4293

Wu ACM, Bough WA, Holmes MR, Perkins BE (1978) Biotech Bioeng 20:1957–1968

Wu H, Wan Y, Dalai S, Zhang R (2010) Response of rat osteoblasts to polycaprolactone/chitosan blend porous scaffolds. J Biomed Mater Res A 92:238–245

Xie H, Li S, Zhang S (2005) Ionic liquids as novel solvents for the dissolution and blending of wool keratin fibres. Green Chem 7:606–608

Yamouchi K, Mniwa M, Mori T (1998) J Biomater Sci Polym Edn 9:259–270

Yin J, Rastogi S, Terry A, Popescu C (2007) Biomacromolecules 8:800–806

Yin J, Rastogi S, et al (2007) Biomacromolecules 8: 800–806

Zeng CH, Qi L (2011) J Text Res 32:12

Zhang Y, Zhang M (2001) Synthesis and characterization of macroporous chitosan/calcium phosphate composite scaffolds for tissue engineering. J Biomed Mater Res 55:304

Zhou Y, Yang D, Chen X, Xu Q, Lu F et al (2008) Electrospun water-soluble carboxyethyl chitosan/poly (vinyl alcohol) nanofibrous membrane as potential wound dressing for skin regeneration. Biomacromolecules 9:349–354

Ziabicki A (1967) Physical fundamentals of the fibre-spinning processes. In Mark HF, Atlas SM (eds) Man–made fibers. Interscience Publishers, New York

Fiber Extraction from Okra Plant Agricultural Wastes, Their Characterizations and Surface Modifications by Environmental Methods

Emine Dilara Kocak, Nigar Merdan, Ilker Mistik and Burcu Yılmaz Sahinbaskan

Abstract Scientists have been searching for biodegradable natural materials that can be used in place of synthetic materials that have not been biodegradable for a long time. Especially in automotive, space, furniture, construction, medical and packaging industries, synthetic composite materials are used abundantly. For this reason, interest in biodegradable biocomposites is increasing. In recent years, many stem fibers such as flax, hemp, kenaf and jute have been started to be used as an alternative to reinforcing fibers traditionally used in composite materials. In fact, researches have begun to focus on the assessment of agricultural plant stem wastes in fiber extraction because of their sustainable, recyclable, biodegradable, renewable and economical properties. Stem waste fibers can show different characteristics according to the plant from which they are obtained. Okra is an agricultural plant that is easy and effortless to cultivate because of its drought-resistant nature and low water requirements. The okra fibers are obtained from the stem wastes of the okra plant (*Abelmoschus esculentus*) remaining on the fields after harvest. The okra fibers have a low ratio of lignin (7.1%), which causes yellowing and photochemical degradation, and have high molecular weight. For this reason, properties such as color fastness and strength are good. The physical and chemical properties of the okra fibers, which have a high cellulose content (67.5%), resemble other traditional body fibers. In terms of usability in the production of composites, cellulose is the most important content in natural fibers. The higher the cellulosic ratio, the stronger the fiber, and so the more suitable to use it as a reinforcement. Recent studies have shown that the mechanical strength and modulus of okra fibers are good and that they have the potential to be used as a reinforcing element in polymer matrix composites. Okra fibers can be modified by environmentally friendly chemical surface modifications, besides improving their

E. D. Kocak (✉) · I. Mistik · B. Y. Sahinbaskan
Faculty Technology Textile Engineering Department, Marmara University, Istanbul, Turkey
e-mail: dkocak@marmara.edu.tr

I. Mistik
e-mail: imistik@marmara.edu.tr

N. Merdan
Architecture and Design Faculty, Istanbul Commercial University, Istanbul, Turkey

S. S. Muthu (ed.), *Sustainable Innovations in Textile Fibres*, Textile Science and Clothing Technology, https://doi.org/10.1007/978-981-10-8578-9_3

mechanical properties such as strength, these treatments increase their absorbency in subsequent processing by providing surface roughness.

Keywords Okra fibers · Surface modifications · Environmental methods

1 Introduction

In recent years scientists are focused on biodegradable natural materials which can be used in stead of synthetic materials. Synthetic composite materials are widely used in automotive, space, construction, medical and packaging industries. So, the interest on biodegradable biocomposites is increasing.

Stem fibers such as hemp, flax, kenaf and jute are used as reinforcement material in composite structures. Also researches are fronted to using of agricultural waste plant stems as source of fiber due to their recyclable, bio-degradable and economic properties. Stem fibers are featured different characteristics due to the plant extracted. Therefore, the most important issue is morphological, thermal and mechanical properties of stem fibers which enables stem fibers to be used as reinforcement material in polymer composites.

Okra plant is grown easily due to the resistant to drought and it requires less water. Okra is used as food for centuries. But researches on utilising stem wastes have been performing for 50 years (Franklin 1982). Researches on extracting fiber from stem waste have been carrying out for 10 years (Alam and Khan 2007). Okra fibers are extracted from waste of okra plant (*Abelmoschus esculentus*) which belongs to malvaceae family. Okra stem fibers contain 67.5% cellulose, 15.4% hemicellulose, 7.1% lignin, 3.4% pectin, 3.9% oil-vax and 2.7% water-soluble substances (Alam and Khan 2007). Lignin content of okra fiber which causes yellowing and photochemical decomposition is low. And it has high molecular weight. Therefore colour fastness and mechanical properties of the okra fibers are good (Kumar et al. 2013). Properties of the okra fibers and other traditional stem fibers are similar and okra fiber has bright and strong structure.

The most important content of the natural fiber is cellulose in terms of using in the composite production. Increasing of cellulose content makes the fiber more stronger and it makes the fiber more appropriate for using as reinforcement material in composites (Mwaikambo and Ansell 2002). Hydroxyl groups (–OH) which presence in the structure of cellulose, hemicellulose and lignin, bond between the macromolecules on the cell wall of the fiber. Exposure to the moisture of the fibers causes breakage of the bonds and hydroxyl groups make new bonds with water molecules, and this causes swelling of the fiber. Therefore, different surface treatments are applied to the natural fibers to swell the cell walls and enable penetration of big chemical molecules into the crystalline zones of the fibers (Mwaikambo and Ansell 2002). Amorphous zones of the cellulose easily absorb the dyes and resins, but compact structure of the crystalline zones make it difficult (Khan and Alam 2013). Surface treatments of the natural fibers help to remove the impurities as well as swell the crystalline zones and remove the hydrophilic hydroxyl groups. Reinforcing of the composite by the natural fiber depends on moisture content of the fiber, fiber-matrix adhesion, crystalline zone and cellulose content of the fiber (Mwaikambo and Ansell 2002).

In recent years, thermal and mechanical properties of the okra fibers were investigated by the researchers, good mechanical and modulus results were obtained and it was reported that okra fibers can be used as reinforcement material in polymer composites (Kumar et al. 2013; Mwaikambo and Ansell 2002; Khan and Alam 2013). Different properties of the okra fibers were improved by chemical (Kumar et al. 2013; Mwaikambo and Ansell 2002; Khan and Alam 2013; De Rosa et al. 2010) and enzymatic surface treatments (Yılmaz et al. 2016).

Also FTIR observations show that okra fibers have similar chemical and physical properties with other traditional lignocellulosic fibers (Kumar et al. 2013; Mwaikambo and Ansell 2002). In order to use okra fibers in composite structures, surface treatments have high importance. Application of alkali surface treatment to okra fibers, improved the properties of composite laminates (De Rosa et al. 2010). Also surface treatment of okra fibers with alkali, improved the crystallization properties of the PLA polymer (Fortunati et al. 2013). It is reported that okra fiber reinforced polyester composites has good dielectric and insulation properties.

Cellulosic fibers are natural, recyclable, renewable, biodegradable and sustainable materials. But they are hydrophilic materials, absorb the moisture easily and that cause the degradation. Surface modification processes are applied to the cellulosic fibers to increase the industrial usage of the fibers. Application of surface treatment processes improve the mechanical properties of the fibers, enable the surface roughness and increase the absorbency of the fibers. And many chemicals can be used for that purpose, such as sodium hydroxide (NaOH) (Khan and Alam 2013; De Rosa et al. 2010; Yılmaz et al. 2016; Arifuzzaman Khan et al. 2009; De Rosa et al. 2011; Moniruzzaman et al. 2009; Onyedum et al. 2016; Srinivasababu 2015), or sodium sulfate (Na_2SO_4) (De Rosa et al. 2011; Fortunati et al. 2013) for alkali process, sodyum hypochlorite (NaClO) or sodium chlorite ($NaClO_2$) (Khan and Alam 2013; Fortunati et al. 2013; Arifuzzaman Khan et al. 2009; Moniruzzaman et al. 2009) for bleaching, acetic acid for acetylation (Khan and Alam 2013; De Rosa et al. 2011; Onyedum et al. 2016).

Surface treatment processes widely remove the lignin, hemicellulose and pectin so wider reaction zones occur on the structure of the cellulose (Onyedum et al. 2016). Concentration of the chemical is an important factor for the surface treatment of the fiber, when it is increased mechanical properties of the fibers decrease (Srinivasababu 2015).

Cellulose content of the okra stem fiber reaches to 75–80% when the alkali surface treatment process is applied due to the removing of hemicellulose, pectin and water soluble substances (Arifuzzaman Khan et al. 2014).

Sodium hydroxide is the mostly used chemical for alkalization. By using alkalization process cellulose I turns to cellulose II. Alkalization, depolarized the molecular structure of the cellulose I as short crystallites (Mwaikambo and Ansell 2002).

Table 1 Properties of the okra stem obtained from Marmara zone

Properties of the okra stem	
Length (cm)	~120
Diameter (cm)	4.3

After bleaching process with $NaClO_2$ roughness of the fiber surface is increased just like alkali process. After bleaching process fibers become softer. Also cellulose content and crystallinity of the fiber is increased due to the removing of lignin and impurities (Khan and Alam 2013).

2 Classification of the Okra Types Grown in Different Zones of the Turkey and Harvesting the Stem Wastes

36,000 ton/year okra plant is grown in Turkey and 211 ton/year plant stem waste is remained on fields (Srinivasababu et al. 2009).

These remained stem wastes are generally burnt after harvesting. Okra plant is grown on Aegean, Black sea, Marmara and Mediterranean zones of the Turkey. Stem thickness of the okra plants are differed due to the okra plant type and the zone. In this study okra stem wastes are obtained from Marmara Zone (Balikesir) and classified according to their stem length and diameter (Table 1).

3 Extraction of Fibers from Okra Stem Wastes

- Okra stem wastes obtained from Marmara zone classified and separated due to their length and diameter.
- Okra stem fibers were extracted form the stem of the plant by using specially designed machine which is used for the extraction of lignocellulose fibers of Turkey (Fig. 1). Fiber extraction machine was registered to Turkish Patent Institute as utility model (Registration number: 2010 08487) in 2010 by the project coordinator and project researcher. This machine is located in Marmara University Faculty of Technology Department of Textile Engineering Physical Testing Laboratory. This lignocellulosic fiber extraction machine has portable, economic, safe, easy-to-use and transportable design. Its efficiency is 100 kg/hour. It is possible to process different fiber types used in textile or industrial fields by changing the method, speed and wires on the machine.
- In order to remove the sticky impurities remained on the extracted fiber, they were soaked in a container for 20 days.
- After 20 days fibers were rinsed by running water and dried in Binder brand drying-oven at 100 °C for 2 h (Fig. 2).

Fig. 1 Okra stem fiber extraction process and extracted okra stem fibers

Fig. 2 Drying-oven, drying process of fibers and dried okra stem fibers

- Dried fibers were conditioned for 24 h under laboratory conditions (65% ± 2 relative humidity and 20 ± 2 °C).

4 Application of Surface Treatment by Different Methods to Extracted Okra Stem Fibers

Extracted okra stem fibers are amorphous due to their lignocellulosic base and the most important contents are cellulose, hemicellulose and lignin (Fig. 3). Characteristic properties of each content have important role on the properties of the whole fiber. Okra stem fibers are bonded to each other because of the lignin, gum and wax. After extracting of fibers hydrophilic process should be applied to remove water-repellent substances such as lignin etc.

3 different methods were used for the surface treatment processes okra stem fibers (conventional method, ultrasonic energy method and microwave energy method). In literature there was no research on surface treatment of okra stem fibers by using ultrasonic and microwave energy methods. Both methods are ecological methods

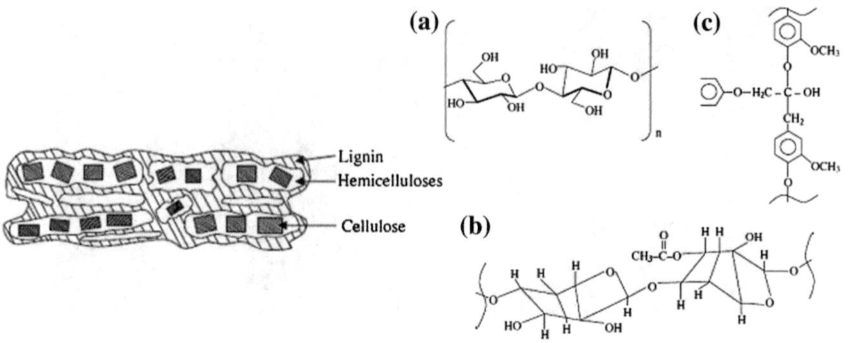

Fig. 3 Structural organization and chemical structure of 3 main components on the cell wall, cellulose **a**, hemicellulose **b**, lignin **c** (Arifuzzaman Khan et al. 2009)

Table 2 Parameters of the sodium hydroxide surface treatment processes

Methods	Duration (min)	Temperature (°C)	Concentration (%)
Conventional	10, 20, 30, 40	60	3, 5, 7, 10
Ultrasonic energy	5, 10, 15, 20	60	3, 5, 7, 10
Microwave energy	3, 5, 7, 10	60	3, 5, 7, 10

which provide water, energy and time savings. Parameters of the 3 different surface treatment methods are given in Table 2.

Energy consumption of conventional, ultrasonic energy and microwave energy methods were calculated according to Formula 1. So energy savings of the 3 surface treatment methods were determined (Table 3).

Table 3 Energy consumption of okra stem fiber surface treatment methods

Methods	Process duration (min)	Power consumed by instrument (W/saat)
Conventional	10	3000
	20	3000
	30	3000
	40	3000
Microwave energy	3	800
	5	800
	7	800
	10	800
Ultrasonic energy	5	660
	10	660
	15	660
	20	660

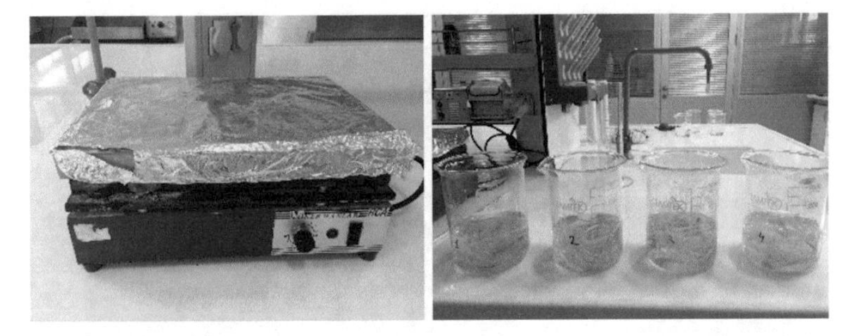

Fig. 4 Heater and processing of the okra stem fiber samples in heated solution for the conventional method

$$P = mCp.\Delta T/t \quad (1)$$

P = Power (j/s), *m* = *Consumed water* (g), C*p* = *Specific heat* (4.18 J/g/K), ΔT = K (*First temperature—last temperature*), *t* = second (s).

Surface treatment processes were performed in 4 different concentration and duration;

- For conventional atmospheric method, it was performed by using Kermanlar Laboratory sample heater at 3000 watts and 220 volts,
- For ultrasonic energy method, it was performed by using Alex ultrasonic washing bath at 220 volts, 660 watts and 40 kHz,
- For microwave energy method, it was performed by using Kenwood Mw440 microwave oven at 300 watts and set to 'Low' setting.

After surface treatment processes, okra stem fibers were rinsed by distilled water (24 °C, pH 7) for 2 min then dried at 100 °C for 2 h by using drying-oven.

In this study alkali (sodium hydroxide) surface treatment processes were performed by using conventional, ultrasonic energy and microwave energy methods. Sodium hydroxide was used in 4 different concentrations and 4 different time settings and same temperature, these procedures were applied according to Islam and Pickering's study (Islam and Pickering 2014).

Application steps of the conventional method are given below:

1. Fibers were weighed by using precision balance and 2 g fiber samples were prepared.
2. 1:20 ratio 400 ml distilled water was heated to 60 °C in Kermanlar Laboratory sample heater (Fig. 4).
3. Chemical substances were added to the bath on appropriate concentrations.
4. 2 g fiber samples were added to the bath and processed for the planned duration (Fig. 4).
5. At the end of the time, fibers were taken out from the bath and rinsed with distilled water for 4–5 times.

Fig. 5 Preparation of the solution—okra stem fiber sample for the microwave energy method and used microwave oven

6. Sodium hydroxide (Merck) treated samples were soaked in 5% acetic acid solution for 5 min for the neutralization.
7. Then fiber samples were taken out and rinsed with distilled water for 4–5 times again.
8. Fiber samples were dried in drying-oven at 100 °C for 2 h.

Application steps of the microwave energy method are given below:

1. Fibers were weighed by using precision balance and 2 g fiber samples were prepared.
2. Chemical substances were added to 1:20 ratio 400 ml distilled water on appropriate concentrations.
3. 2 g fiber samples were added to the solution.
4. Prepared samples were put in Kenwood Mw440 microwave oven, then samples were processed in microwave oven for planned duration at 'Low' energy setting (Fig. 5).
5. At the end of the duration fiber samples were taken out from the solution and rinsed with distilled water for 4–5 times.
6. Sodium hydroxide (Merck) treated samples were soaked in 5% acetic acid solution for 5 min for the neutralization.
7. Then fiber samples were taken out and rinsed with distilled water for 4–5 times again.
8. Fiber samples were dried in drying-oven at 100 °C for 2 h.

Application steps of the ultrasonic energy method are given below:

1. Fibers were weighed by using precision balance and 2 g fiber samples were prepared.
2. Chemical substances were added to 1:20 ratio 400 ml distilled water on appropriate concentrations.
3. 2 g fiber samples were added to the solution in the ultrasonic bath.

Fig. 6 Ultrasonic energy bath and processing of the okra stem fibers for the ultrasonic energy method

4. Samples were processed in Alex ultrasonic bath (220 volts 660 watts 40 kHz) for the planned duration (Fig. 6).
5. At the end of the duration fiber samples were taken out from the solution and rinsed with distilled water for 4–5 times.
6. Sodium hydroxide (Merck) treated samples were soaked in 5% acetic acid solution for 5 min for the neutralization.
7. Then fiber samples were taken out and rinsed with distilled water for 4–5 times again.
8. Fiber samples were dried in drying-oven at 100 °C for 2 h.

The mostly used surface treatment method is alkali treatment. Type of the chemicals, concentration, process duration and temperature affect the alkali treatment. By using alkali treatment lignin and hemicellulose of the fiber are removed. In this study, during the alkali surface treatment process, OH groups are reacted with the water molecules (Formula 2), so H-bonds are broken, amorphous structure and fibrillation of the fiber are increased (Fig. 7). By breaking H-bonds surface unevenness is increased, this enables better adhesion between fiber and matrix, so strength of the composite is increased (Hamideh et al. 2014).

$$\text{Fiber}{-}\text{OH} + \text{NaOH} \rightarrow \text{Fiber}{-}\text{O}{-}\text{Na} + H_2\text{O} \quad (2)$$

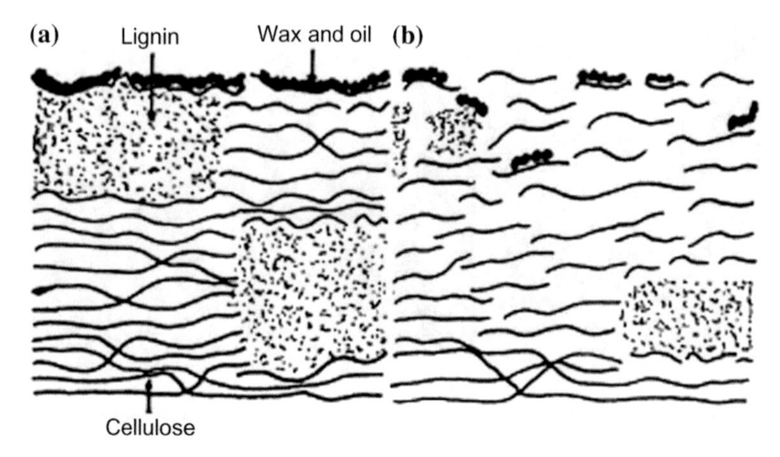

Fig. 7 Structure of the alkali treated (**a**) and untreated (**b**) fiber (Mwaikambo and Ansell 2002)

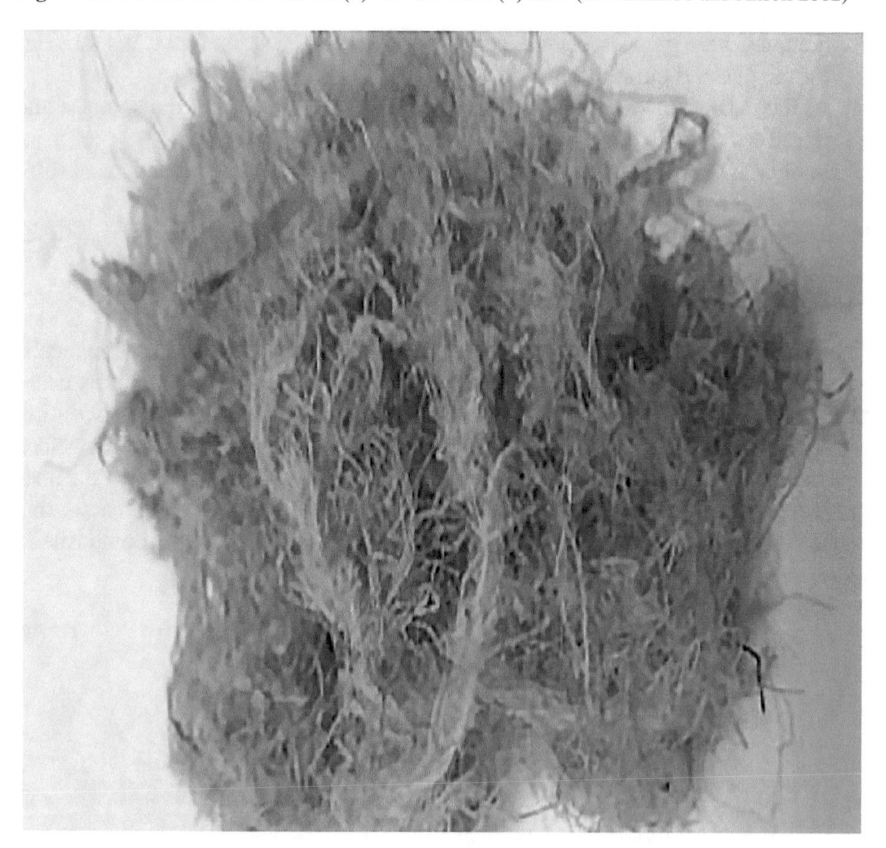

Fig. 8 Reaction between okra stem fiber and sodium hyroxide

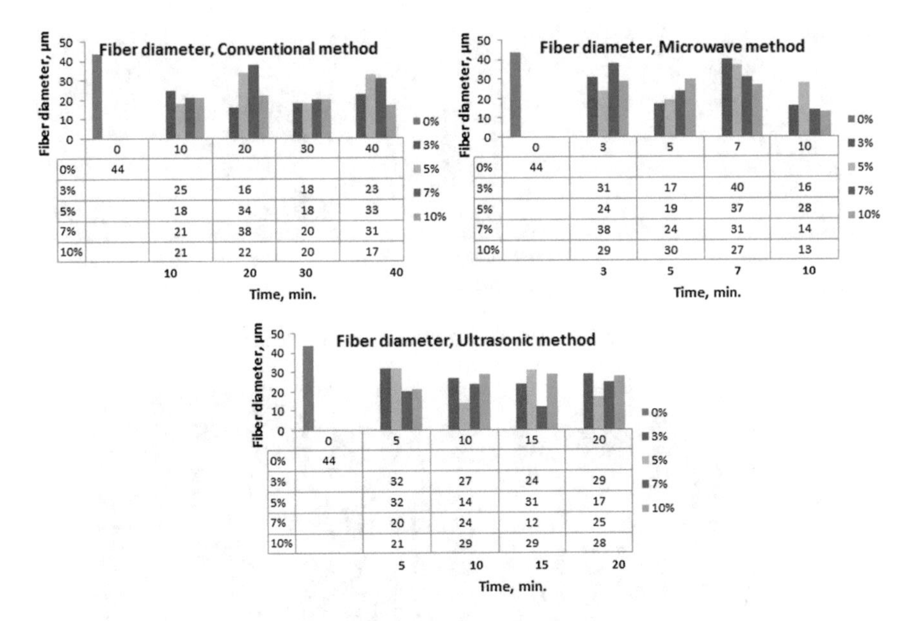

Fig. 9 Fiber diameters of untreated and treated okra stem fibers

5 Analysing of Okra Stem Waste Fibers

5.1 Measurement of Diameter and Observation of Longitudinal and Cross-Sectional Views of Okra Stem Fibers

Stem diameter of the okra plant which grown in Turkey changes between 10 and 50 mm. Longitudinal and cross-sectional views and diameters of the treated and untreated okra stem fibers were determined in Marmara University Technology Faculty Department of Textile Engineering Physical Testing Laboratory by using Projectina CH-9495 projection microscope and Olympus CH_2 light microscope (Figs. 8, 9 and 10).

It was observed that brightness and touching properties of the sodium hydroxide treated okra stem fibers were increased.

After surface treatment process, okra stem fibers swelled and their cross-sections extended. But during the surface treatment process hemicellulose, lignin and other impurities were removed therefore cross-section of the fibers were slightly reduced.

Longitudinal views and diameter measurements of the okra stem fibers are given in Fig. 10.

When longitudinal views and diameters of the okra stem fibers were investigated, linear deformation was observed same as other lignocellulosic fibers (De Rosa et al. 2010).

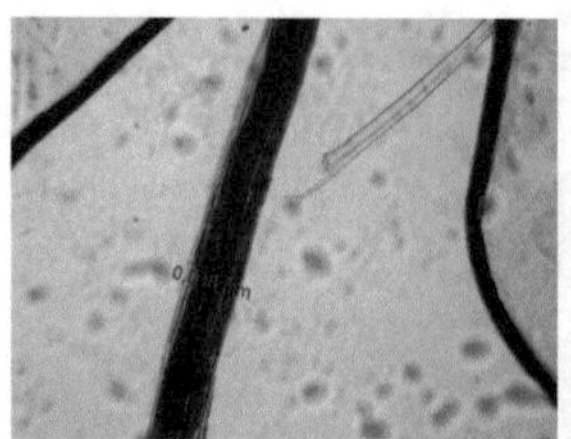

Conventional %7 Sodium hydroxide for 20 min

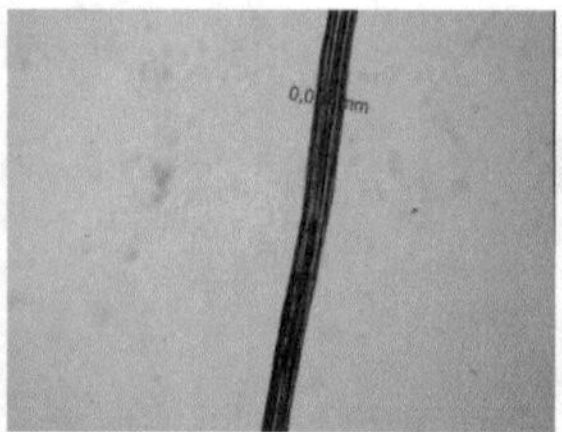

Microwave %10 Sodium hydroxide for 3 min

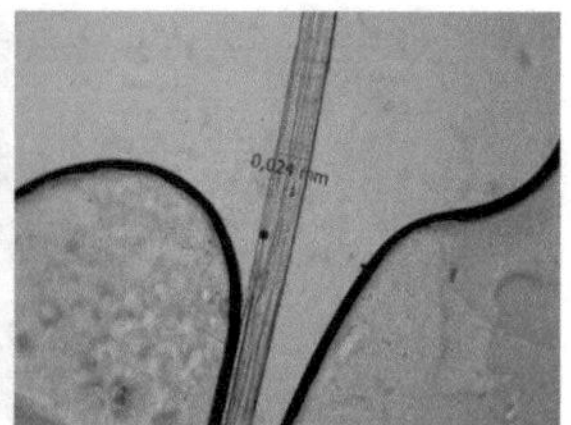

Ultrasonic %5 Sodium hydroxide for 15 min

Fig. 10 Longitudinal views and diameter measurements of okra stem fibers

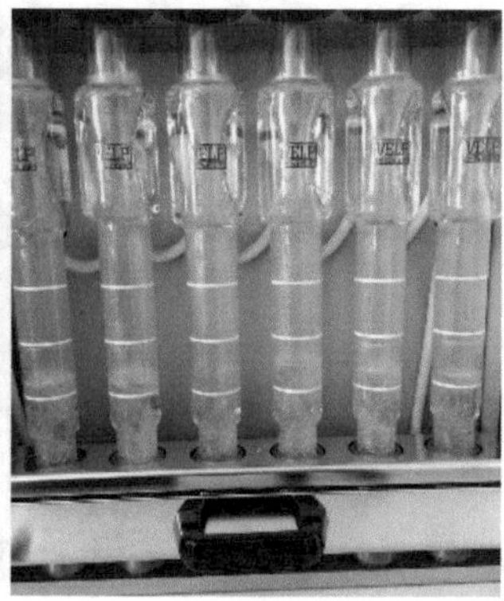

Fig. 11 Cellulose determination instrument and processing of okra stem fibers

Swelling is a physicochemical occurence happens when cellulosic fibers meet with water and chemicals, water molecules penetrate to the fiber and breaks the hydrogen bonds. Thus cellulosic fibers become appropriate for swelling. So water molecules bond with the OH– groups of cellulose fiber, and then cellulose fibers start to swell. The degree of the swelling depends on the number of the free hydroxyl (OH–) groups (Bledzki and Gassan 1999; Gassan and Bledzki 1998).

5.2 Determination and Evaluation of Cellulose, Hemicellulose and Lignin Amount of the Okra Stem Fibers

Cellulose, hemicellulose and lignin amounts of the okra stem fibers were determined by Velp brand FIWE 6 instrument which obtained by TUBITAK 1001 project (Fig. 11). Experiments were performed according to Van Soest and Wine 1968 method (Van Soest and Wine 1968) and AOAC (2016) standard (AOAC Official Methods of Analysis 2016).

Application steps for determination of cellulose ratio:

1. Fiber sample was placed into the crucible and weighed.
2. Crucibles were placed into the instrument.
3. 150 ml 1.25% sulphuric acid solution was added to the tube of the instrument.
4. 10 drops of anti-foam were added to the tubes.
5. The instrument was started and heat increase begun.
6. When the boiling of the water started, the level was set to 5 and process continued for 30 min.
7. At the end of the time, sulphuric acid was vacuumed and distilled water was added.
8. 30 ml boiled water was added to the tube and vacuumed again.
9. 150 ml potassium hydroxide was added to the tube with anti-foam.
10. Instrument was started again and heat increase began.
11. When the boiling of the water started, the level was set to 5 and process continued for 30 min.
12. At the end of the duration vacuuming and cleaning processes were repeated.
13. Samples were taken out from the instrument with crucibles and they were dried in oven at 105 °C for 1 h.
14. After drying, samples were waited for 1 h and then they were weighed.
15. Samples were dried in muffule furnace at 550 °C for 3 h.
16. Samples were weighed after 1-hour wait.
17. Cellulose ratio was calculated according to Formula 3 by using weighing results which were obtained at 3 stages.

$$Cellulose\,ratio\,(\%) = ((F_1 - F_2)\,/\,F_0) \times 100 \quad (3)$$

F_0: First weight (excluding crucible), F_1: Weight after drying (including crucible), F_2: Weight after muffule furnace (including crucible),

Application steps for determination of neutral fiber ratio:

1. Fiber sample was placed into the crucible and weighed (1 g).
2. Crucibles were placed into the instrument.
3. 100 ml neutral solution (sodium boratedehydrate, disodium ethylenediaminetetraacetate, sodiumlaurylsulfate, 2-ethoxyethanol, disodiumphosphateanhydrous) was added to the tube of the instrument.
4. 10 drops of anti-foam were added to the tubes.
5. The instrument was started and heat increase begun.
6. When the boiling of the water started, the level was set to 5 and process continued for 60 min.
7. At the end of the time, neutral solution was vacuumed and distilled water was added.
8. 3 times 30 ml boiled water and 2 times acetones were added to the tube and vacuumed again.

9. Samples were taken out from the instrument with crucibles and they were dried in oven at 105 °C for 8 h.
10. Samples were weighed after 1-hour wait.
11. Neutral fiber ratio (NDF%) was calculated according to Formula 4 by using weighing results.

$$NDF\% \ (neutral\ fiber\ ratio) = ((F_1 - F_k)/F_0) \times 100 \tag{4}$$

F_0: First weight (excluding crucible), F_1: Weight after drying (including crucible), F_k: weight of crucible.

Application steps for determination of hemicellulose ratio:

1. Fiber sample was placed into the crucible and weighed (1 g).
2. Crucibles were placed into the instrument.
3. 100 ml acidic solution ($C_{19}H_{42}BrN + H_2SO_4$) was added to the tube of the instrument.
4. 10 drops of anti-foam were added to the tubes.
5. The instrument was started and heat increase began.
6. When the boiling of the water started, the level was set to 5 and process continued for 60 min.
7. At the end of the duration, acidic solution was vacuumed and distilled water was added.
8. 3 times 30 ml boiled water and 2 times acetones were added to the tubes and vacuumed again.
9. Samples were taken out from the instrument with crucibles and they were dried in oven at 105 °C for 8 h.
10. Samples were weighed after 1-hour wait.
11. Acidic washed fiber ratio (ADF%) was calculated according to Formula 5 by using weighing results.

$$ADF\ \% \ (acidic\ washed\ fiber\ ratio) = ((F_1 - F_k)/F_0) \times 100 \tag{5}$$

F_0: First weight (except crucible), F_1: Weight after drying (including crucible), F_k: Weight of crucible.

Ratio of the hemicellulose is calculated according to Formula 6.

$$Hemicellulose\ \% = NDF\ \% - ADF\ \% \tag{6}$$

Application steps for determination of lignin ratio:

1. Fiber sample was placed into the crucible and weighed (1 g).
2. Crucibles were placed into the instrument.
3. 100 ml acidic solution (72% sulphuric acid) was added to the tube of the instrument.
4. 10 drops of anti-foam were added to the tubes.

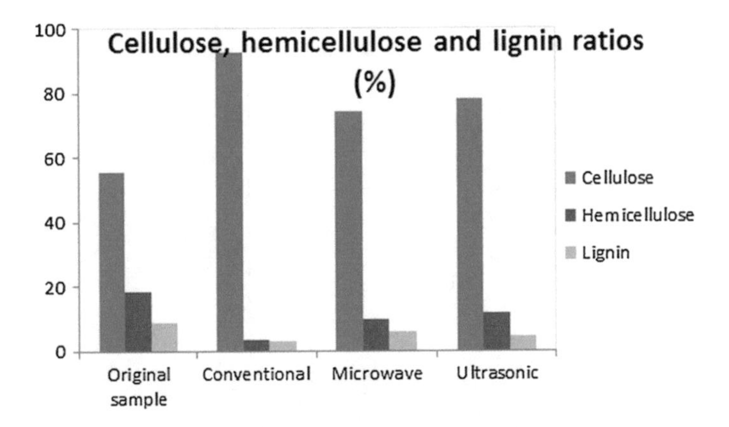

Fig. 12 Cellulose, hemicellulose and lignin ratios of okra stem fibers

5. The instrument was started and heat increase began.
6. When the boiling of the water started, the level was set to 5 and process continued for 60 min.
7. At the end of the duration, acidic solution was vacuumed and distilled water was added.
8. 3 times 30 ml boiled water and 2 times acetones were added to the tubes and vacuumed again.
9. Cold extraction continued for 3 h.
10. At the end of the duration, acidic solution was vacuumed and distilled water was added.
11. 3 times 30 ml boiled water and 2 times acetones were added to the tubes and vacuumed again.
12. Samples were taken out from the instrument with crucibles and they were dried in oven at 105 °C for 8 h.
13. Samples were weighed after 1-hour wait.
14. Lignin ratio (%ADL) was calculated according to Formula 7 by using weighing results.

$$ADL\%\,(acidic\,washed\,lignin\,ratio) = ((F_1 - F_k)\,/F_0) \times 100 \tag{7}$$

F_0: First weight (excluding crucible), F_1: Weight after drying (including crucible), F_k: Weight of the crucible.

Cellulose, hemicellulose and lignin ratios of the untreated and treated okra stem fibers were determined by using cellulose determination instrument. The optimum results (according to time and concentration) were evaluated (Table 4 and Fig. 12).

The highest increase on cellulose ratio after sodium hydroxide treatment was obtained from conventional method (66.7%). When the general chemical structure

Table 4 Cellulose, hemicellulose, lignin ratios and weighing results of chosen samples

Chemical	Methods	Duration (min)	Concentration (%)	First weight of samples (g) (F_0)	Weight of samples after drying (g) (F_1)	Weight of samples after muffule furnace (g) (F_2)	Cellulose ratio (%)	Hemicellulose ratio (%)	Lignin ratio (%)
Control sample				0.054	29.63	29.6	55.55	18.25	8.9
Sodium hydroxide	Conventional	20	7	0.054	29.949	29.89	92.59	3.48	2.87
	Microwave	3	10	0.058	29.605	29.562	74.13	9.85	6.02
	Ultrasonic	15	5	0.055	29.914	29.871	78.18	11.56	4.67

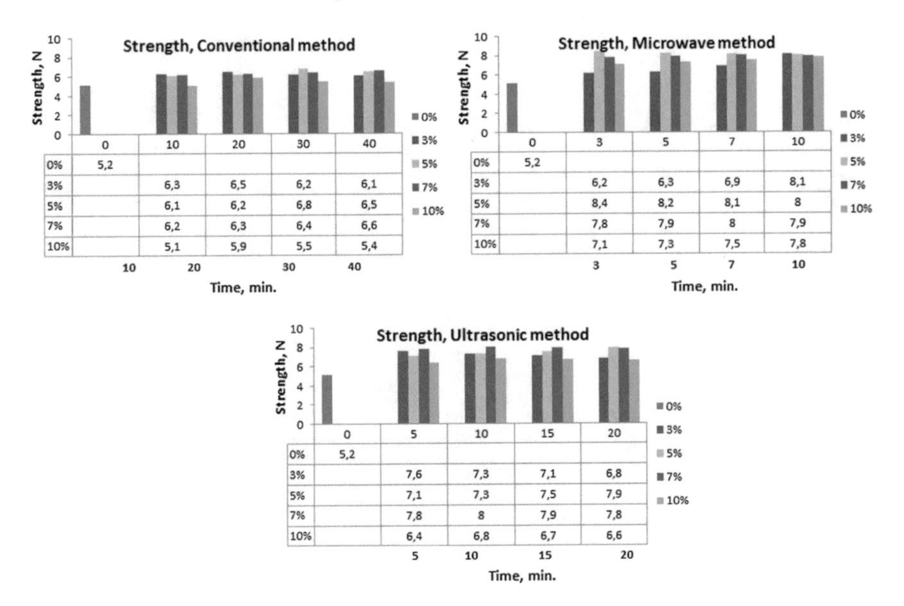

Fig. 13 Tensile strength values okra stem fibers

of the fiber was evaluated, fibrillation and cellulose ratio of the fibers were increased due to the surface treatments.

5.3 *Determination and Evaluation of Tensile Strength, Elongation and Young Modulus of Okra Stem Fibers*

Tensile strength (kgf) and elongation (%) properties of the untreated and treated okra stem fibers were measured according to ASTM D 3822 standard. Testing of the fibers were performed by using Instron 4411 testing instrument (50 N load, 10 mm/min speed) in Marmara University Technology Faculty Department of Textile Engineering Physical Testing Laboratory.

Mechanical properties of the lignocellulosic fibers are affected by morphology, crystallinity, amorphous zone ratio, orientation, physical and chemical properties of the fibers. Moisture absorbtion capacity of the lignocellulosic fibers are differed due to the structure of the fiber. Surface treatment processes are increased the surface roughness and moisture absorbtion capacity of the fiber (Li et al. 2007). Tensile strength values of the untreated and treated okra stem fibers are given in Fig. 13.

Tensile strength values of the okra stem fibers were increased after conventional surface treatment. But after 40 min tensile strength values of the treated fibers were decreased. After surface treatment by microwave energy method again higher tensile

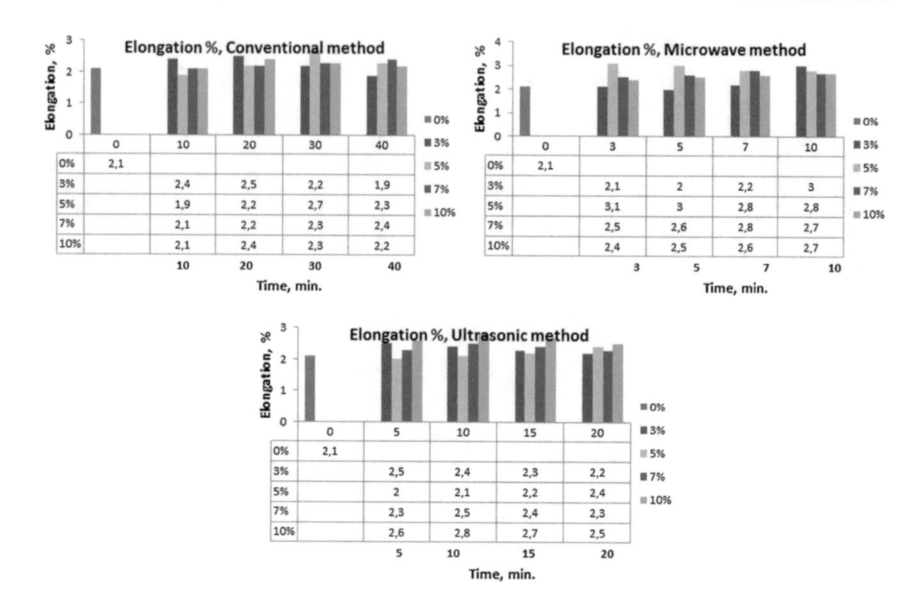

Fig. 14 % elongation values of okra stem fibers

strength values were obtained. The highest tensile strength value was obtained with 5% sodium hydroxide concentration for 5 min process.

Same as conventional and microwave methods, tensile strength values of the okra stem fibers were increased after ultrasonic surface treatment process. But after 20 min tensile strength values of the fibers were decreased (Fig. 13).

Elongation values (%) of the untreated and treated okra stem fibers are given in Fig. 14.

As shown in Fig. 14, the highest elongation values were obtained on 30 min—5% sodium hydroxide concentration, 3 min—5% sodium hydroxide concentration and 10 min—10% sodium hydroxide concentration for conventional, microwave and ultrasonic energy methods respectively.

Young modulus values of the treated and untreated okra stem fibers are given in Fig. 15.

The young modules values in Fig. 15 were over 13 MPa for treated fiber samples on 5 min, 5% sodium hydroxide concentration with microwave energy and on 10 min. 7% sodium hyroxide concentration with ultrasonic process.

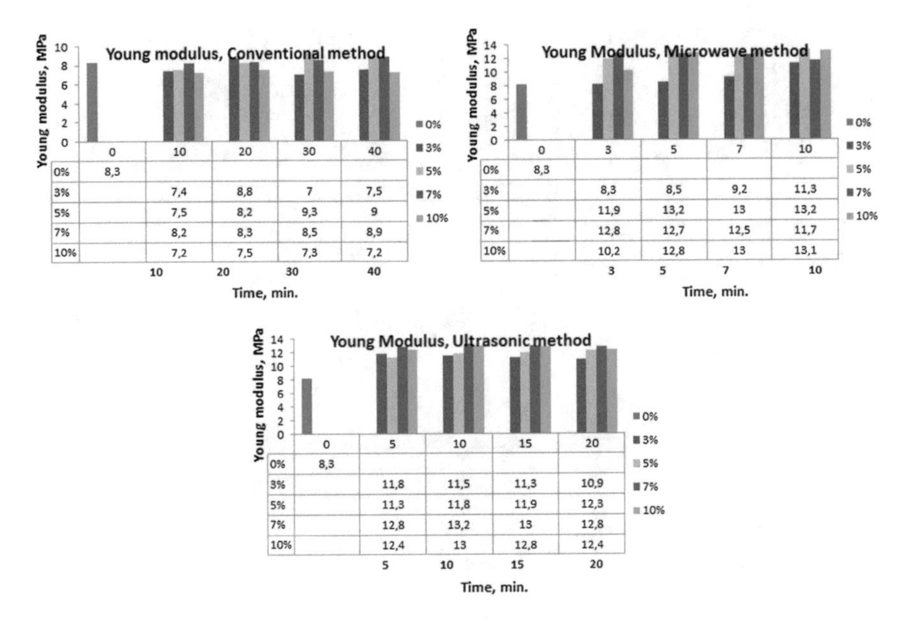

Fig. 15 Young modulus values of okra stem fibers

5.4 *Determination and Evaluation of Weight Loss of Okra Stem Fibers*

Okra stem fibers were conditioned after surface treatment method then they were weighed by precision balance. Weight of each sample before surface treatment was 2 grams. % weight loss of the fibers was calculated according to Formula 8.

$$W_j = ((W_{first} - W_{last}) / W_{first}) \times 100 \tag{8}$$

W_j: Weight loss %, W_{first}: First weight, W_{last}: Weight after treatment

Weight loss of okra stem fibers after conventional, microwave energy and ultrasonic energy surface treatment methods are given in Fig. 16.

The highest weight loss values were obtained on 30 min—5% sodium hydroxide concentration, 3 and 5 min—10% sodium hydroxide concentration and 15 min—5% sodium hydroxide concentration for conventional, microwave energy and ultrasonic energy methods respectively (Fig. 16).

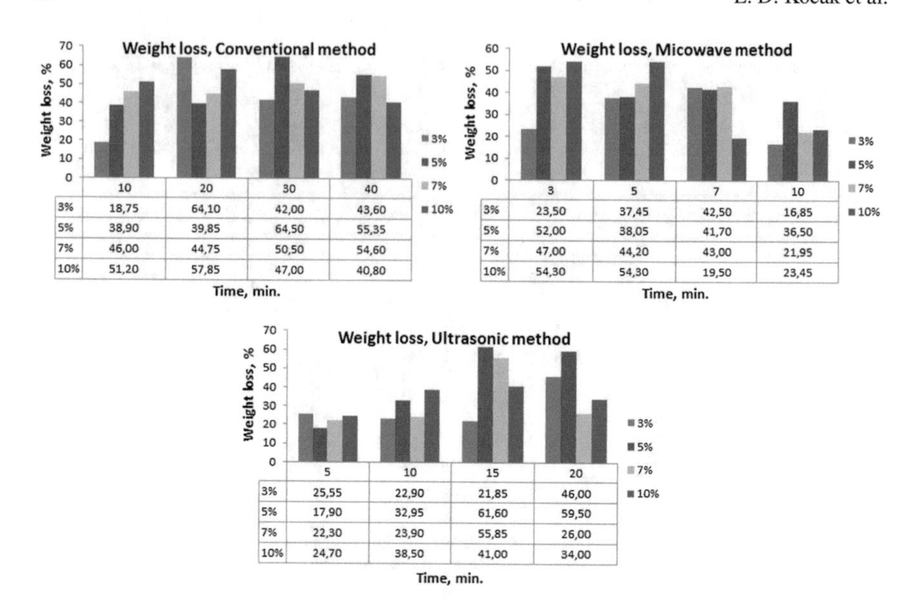

	10	20	30	40
3%	18,75	64,10	42,00	43,60
5%	38,90	39,85	64,50	55,35
7%	46,00	44,75	50,50	54,60
10%	51,20	57,85	47,00	40,80

	3	5	7	10
3%	23,50	37,45	42,50	16,85
5%	52,00	38,05	41,70	36,50
7%	47,00	44,20	43,00	21,95
10%	54,30	54,30	19,50	23,45

	5	10	15	20
3%	25,55	22,90	21,85	46,00
5%	17,90	32,95	61,60	59,50
7%	22,30	23,90	55,85	26,00
10%	24,70	38,50	41,00	34,00

Fig. 16 % weight loss values of okra stem fibers

5.5 *Determination and Evaluation of Chemical Structure (FTIR), Crystallinity (XRD), Thermal (DSC) and Morphological (SEM) Properties of Okra Stem Fibers*

5.5.1 Chemical Structure Analyzing of Okra Stem Fibers (FTIR)

FTIR method was used to investigate the changes on the chemical bonds and effectivity of the surface treatment methods on the okra stem fibers. IR spectrum of the fibers was measured by Perkin Elmer Spectrum 100 FT-IR spectrometer between 4000 and 650 cm^{-1} ranges (resolution: 2 cm^{-1}). The optimum samples were chosen and tested including control sample.

Sodium hydroxide treated okra stem fibers's peak values were investigated up to 4000 cm^{-1} band in Fig. 17. FTIR spectrum of the okra fibers show the absorbtion band of characteristic chemical groups of cellulose, hemicellulose and lignin. Main components are functional groups which include alkine, aromatic groups and oxygen (ester, ketone and alcohol). Wide absorbtion peak between 2930 and 3330 cm^{-1} band corresponds to typical O–H stretching and H bonds of hydroxyl groups. Band range between 2930 and 2850 cm^{-1} corresponds to C–H stretching of cellulose and hemicellulose. Absorbance at 1731 cm^{-1} band belongs to C=O stretching due to the decreasing of ester groups in hemicellulose or decreasing of carboxyl in lignin and disappeared with the sodium hydroxide treatment. The reason of that removing of carboxyl or carbonyl groups. The peak between 1520 and 1590 cm^{-1} band corresponds to C=C stretching of aromatic rings in lignin. Decrease at 1430 cm^{-1} band

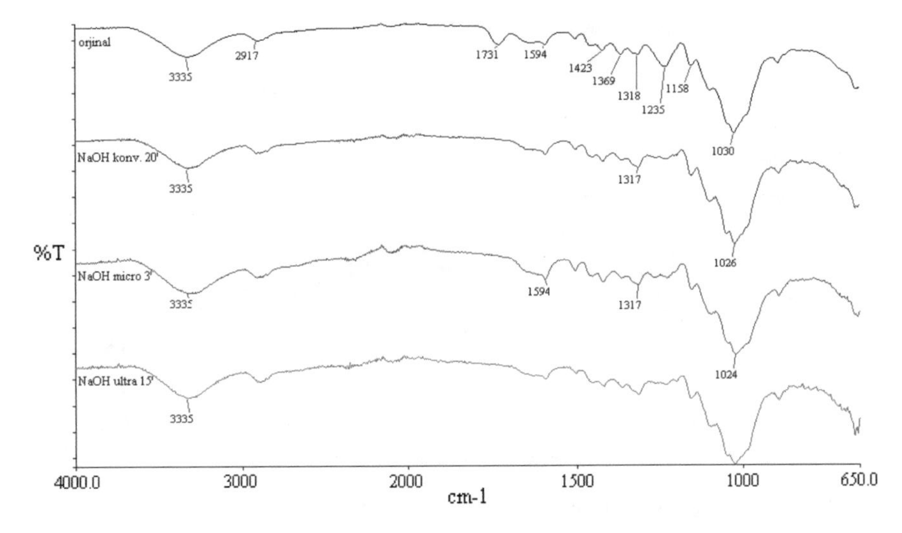

Fig. 17 FTIR spectrum of sodium hydroxide surface treated okra stem fibers

belongs to CH_2 symmetric inflection. The peaks at 1390 cm^{-1} and 1260 cm^{-1} correspond to acetyl groups of lignin and C–O stretching of hemicellulose respectively. The peaks at 1350 and 1320 cm^{-1} belongs to aromatic rings of polysaccharide (C–H) and inflection vibrations of C–O groups (Fortunati et al. 2013). The absorbance peak at 1170 cm^{-1} corresponds to anti-symmetric deformation of C–O–C band. All these results show that crystalline structure of the fibers changed from cellulose I form to cellulose II form after sodium hydroxide treatment (Arifuzzaman Khan et al. 2009).

5.5.2 Crystallinity of Treated and Untreated Okra Stem Fibers (XRD)

Crystallinity of the treated and untreated okra stem fibers was measured by D8 Bruker aXS Advance X-ray diffractometer at 40 kV and 40 mA in Marmara University Engineering Faculty Department of Environmental Engineering. Diffraction scanning limit was 2°/s and measurement range was 0°–50°. Crystallinity index was calculated according to Formula 9 from the measurement results.

$$Crystallinty\,index\,(\%) = ((I_{002} - I_{am})\,/\,I_{002}) \times 100 \quad (9)$$

I_{002} Maximum 002 cage reflection intensity at $2\theta = 22°$ of cellulose crystallographic material.

I_{am} Diffraction intensity of amorphous material at $2\theta = \sim 18°$

Crystallinity index results of untreated and sodium hydroxide treated okra stem fibers are given in Figs. 18, 19 and 20.

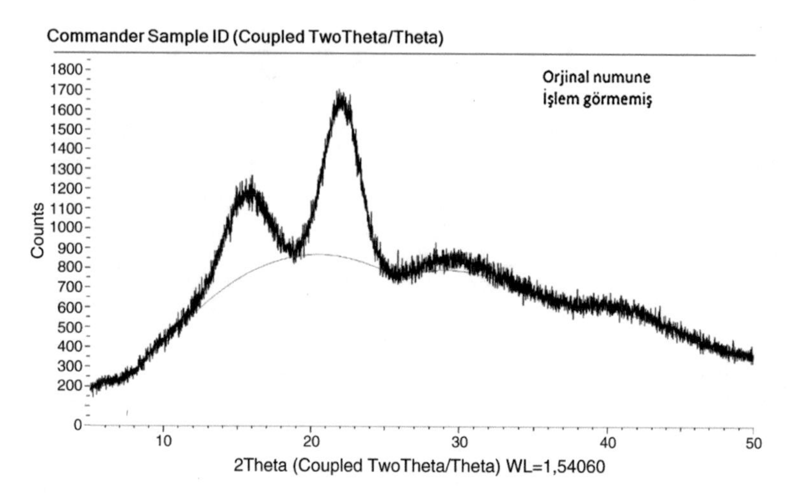

Fig. 18 X-ray diffractometer graphic of untreated okra stem fibers

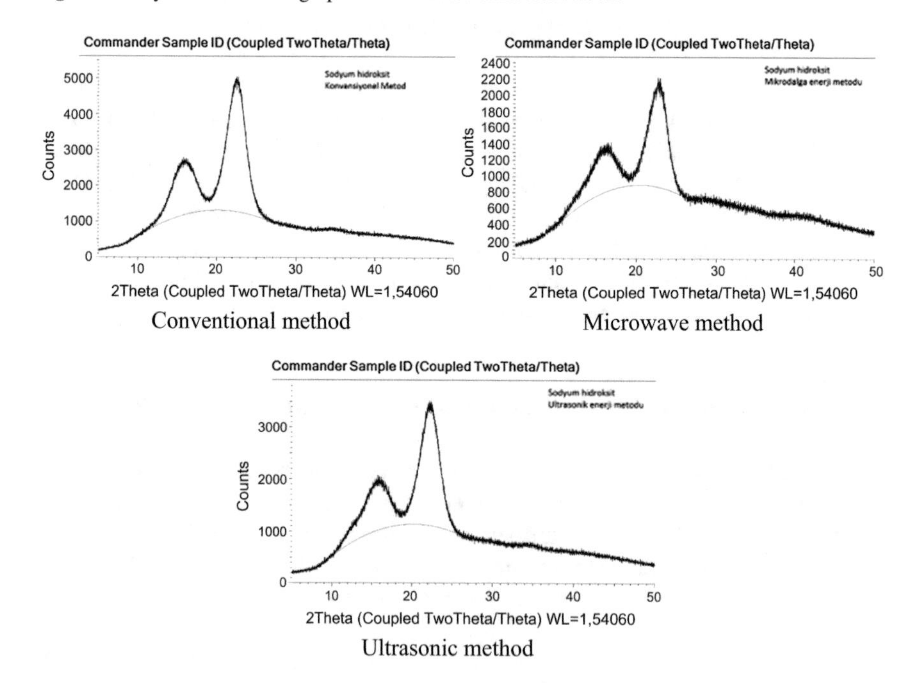

Fig. 19 X-ray diffractometer graphics of surface treated okra stem fibers

Crystallinity index values of the conventional sodium hydroxide treated samples were decreased by increasing the concentration. The highest result was obtained from 30 min—5% sodium hydroxide treatment. Increasing of duration and concentration generally decreased the crystallinity index values (Fig. 20).

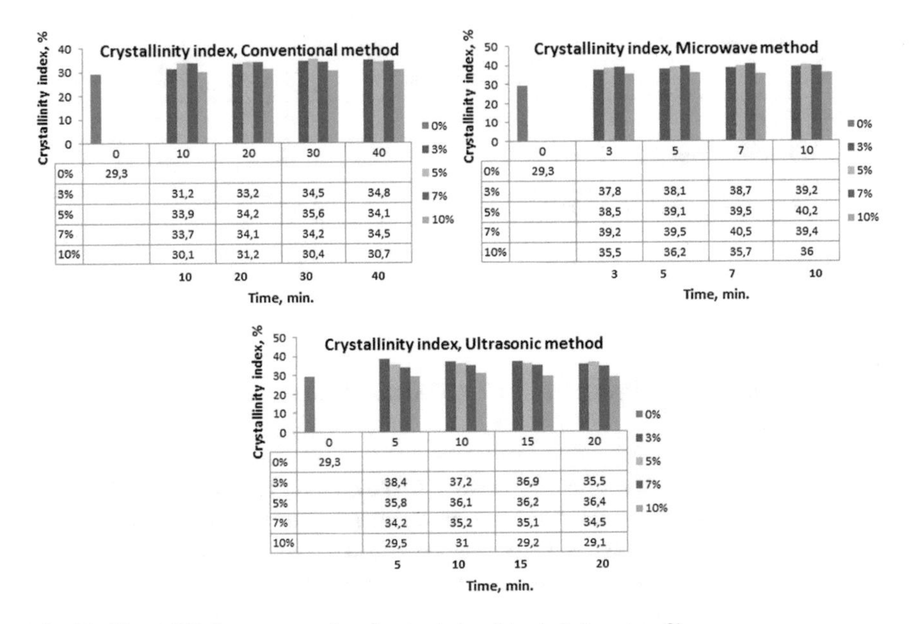

Fig. 20 X-ray diffractometer results of untreated and treated okra stem fibers

In microwave energy method, crystallinity index values generally increased by increasing of the process duration. The highest result was obtained from 10 min—5% sodium hydroxide treatment. Increasing of concentration decreased the crystallinity index values (Fig. 20).

In ultrasonic energy method, crystallinity index values decreased by increasing of the concentration. The highest result was obtained from 5 min—3% sodium hydroxide treatment (Fig. 20).

5.5.3 Thermal Properties of Untreated and Treated Okra Stem Fibers (DSC)

Thermal properties of the untreated and treated okra stem fibers were determinated by using Perkin Emler brand DSC instrument in Istanbul Textile and Raw Materials Exporters Union, Istanbul Textile Research and Development Centre Laboratory. Thermal analyses were performed between 0 and 350 °C, at the speed of 10 °C/min under Nitrogen atmosphere. DSC thermographs and DSC values are given in Fig. 21 and Table 5 respectively.

When the DSC thermographs of treated and untreated okra stem fibers were investigated, only minor differences were observed on melting points. But enthalpy value of ultrasonic treated okra stem fibers increased by 19.2%. The reason of that is the effect of ultrasonic energy on -OH groups of -COOH groups of okra stem fibers. This effect caused an increase on enthalpy values.

Table 5 DSC results of the samples

Samples	T_m(°C) melting point	ΔH melting enthalpy (J/g)
Control sample	98.1	20.2
Treated by conventional method	98.7	20.3
Treated by microwave energy method	99.3	20.1
Treated by ultrasonic energy method	101.2	24.2

5.5.4 Morphological Properties of Treated and Untreated Okra Stem Fibers (SEM)

In order to investigate the effect of surface treatments on fibers, longitudinal and cross-sectional morphological properties of the fibers were observed by using JEOL JSM-5410 LV (20 kV) electron microscope (SEM). The surface topography of untreated okra stem fibers and sodium hyroxide treated fibers with conventional, microwave and ultrasonic processes were shown in Figs. 22, 23, 24 and 25, respectively. The porosity of ultrasonic processed fiber surface with sodium hydroxide were more high than the untreated and other processed fibers. When the treated fiber surface with ultrasonic process were considered, it was observed that it was rougher

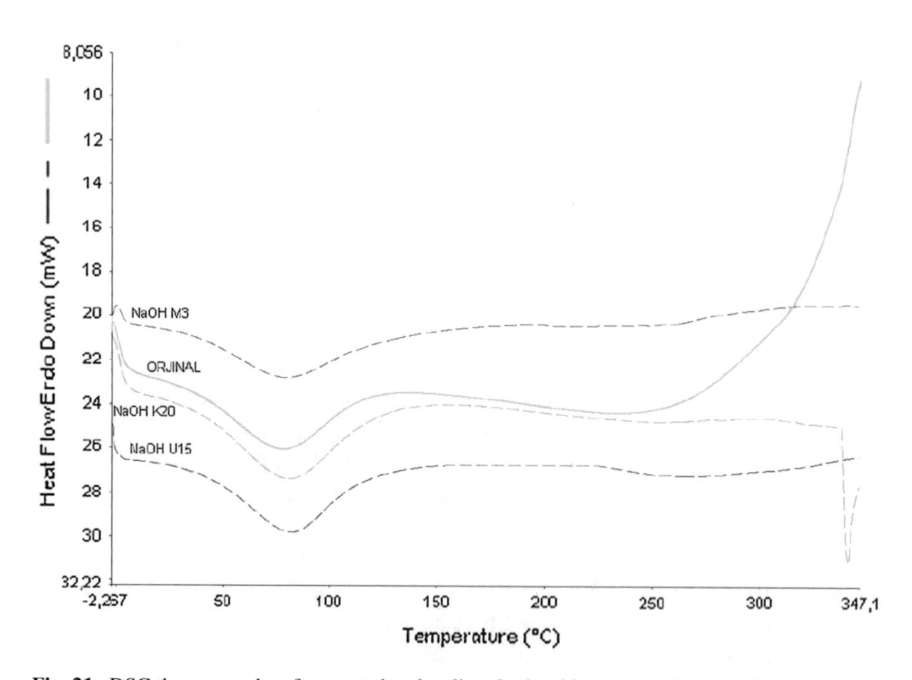

Fig. 21 DSC thermographs of untreated and sodium hydroxide treated okra stem fibers

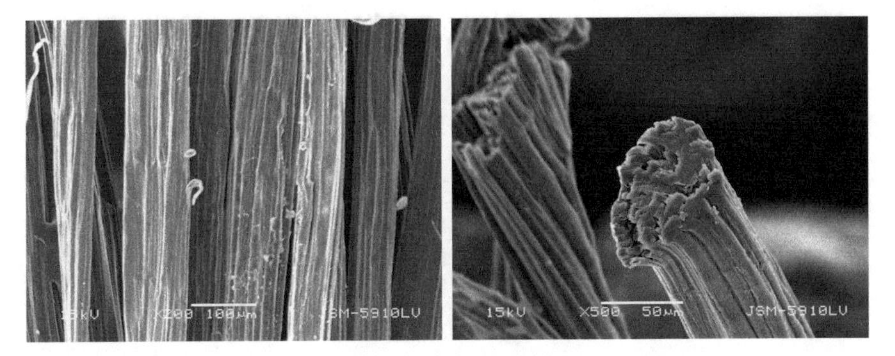

Fig. 22 Longitudinal and cross sectional views of untreated okra stem fibers

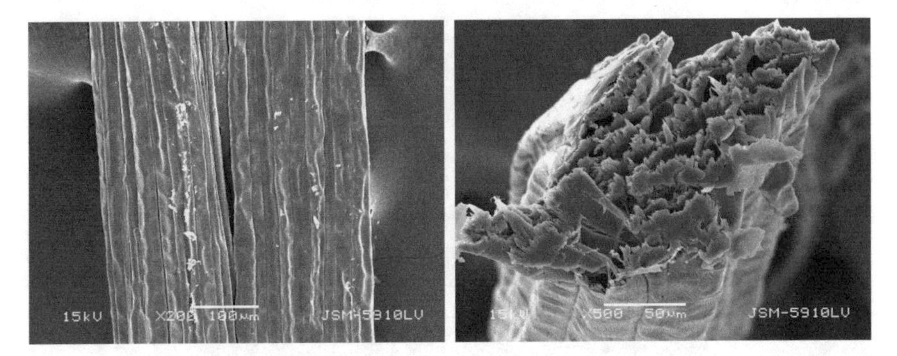

Fig. 23 Longitudinal and cross sectional SEM views of conventional sodium hydroxide treated okra stem fibers

than the untreated and other processed fibers. By the ultrasonic process operations, the extractives, hemicellulose and lignin on the surface were removed. When the morphologies of the untreated and other surface processed fibers were compared, as a result of alkalization realized by the sodium hyroxide operation the hydroxyl groups on the surface. It was observed that it showed cleaning by composing better roughness on the fiber surface.

Surface treatment methods removed the impurities of the okra stem fibers and increased the surface roughness of the okra stem fibers.

5.5.5 Evaluation of Energy Consumption of Conventional, Microwave and Ultrasonic Treatment Processes

Energy consumptions of conventional, microwave and ultrasonic surface treatment methods of okra stem fibers are given in Table 6. Energy consumptions of the surface treatment methods increased according to the duration of the processes. It was observed that the most environment-friendly method was microwave energy method.

Fig. 24 Longitudinal and cross sectional SEM views of microwave sodium hydroxide treated okra stem fibers

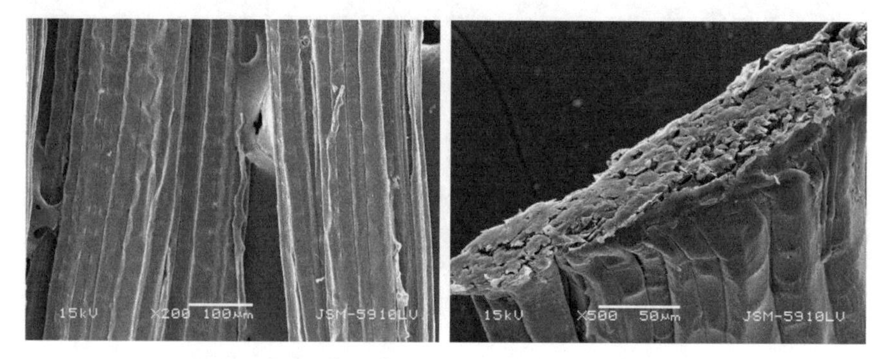

Fig. 25 Longitudinal and cross sectional SEM views of ultrasonic sodium hydroxide treated okra stem fibers

6 Conclusion

Surface treatment methods removed the impurities of the okra stem fibers. Therefore cellulose ratio, tensile strength and elongation properties of the okra stem fibers increased. Higher tensile strength values were obtained from microwave and ultrasonic energy methods. Surface treatment methods also increased the surface roughness of the okra stem fibers that will increase the fiber-matrix adhesion for the production of the composites. Crystallinity index values of the fibers increased after surface treatments. Surface treatment methods have only minor effect on melting point of the fibers. By using microwave and ultrasonic energy methods less energy was consumed. As a result by using microwave and ultrasonic energy methods generally more effective surface treatments can be performed by consuming less energy than conventional methods for okra stem fibers.

Table 6 Energy consumptions of surface treatment methods

Methods	Duration (min)	Energy consumed by the instrument (W/saat)	Energy consumption (W/saat)
Conventional	10	3000	500
	20	3000	1000
	30	3000	1500
	40	3000	2000
Microwave energy	3	800	40
	5	800	66.66
	7	800	93.33
	10	800	133.33
Ultrasonic energy	5	660	55
	10	660	110
	15	660	165
	20	660	220

Acknowledgement This work was supported by TUBITAK 1001 under grant 215M984.

References

Alam M, Khan G (2007) Chemical analysis of okra bast fiber (*Abelmoschus esculentus*) and its physico-chemical properties. J Text Appar Technol 5(4):1–9

AOAC Official Methods of Analysis (2016) [Online]. Available: http://www.aoac.org/aoac_prod_imis/AOAC/Publications/Official_Methods_of_Analysis/AOAC_Member/Pubs/OMA/AOAC_Official_Methods_of_Analysis.aspx?hkey=5142c478-ab50-4856-8939-a7a491756f48. Accessed 13 May 2017

Arifuzzaman Khan G, Ahsanul Haque M, Shamsul Alam M (2014) Studies on okra bast fibre-reinforced phenol formaldehyde resin composites. Biomass and Bioenergy. Springer International Publishing, Springer, pp 157–174

Arifuzzaman Khan GM, Shaheruzzaman M, Rahman MH, Abdur Razzaque SM, Islam MS, Alam MS (2009) Surface modification of okra bast fiber and its physico-chemical characteristics. Fibers Polym 10(1):65–70

Bledzki AK, Gassan J (1999) Composites reinforced with cellulose based fibres. Prog Polym Sci 24:221–274

De Rosa IM, Kenny JM, Maniruzzaman M, Moniruzzaman MD, Monti M, Puglia D, Santulli C, Sarasini F (2011) Effect of chemical treatments on the mechanical and thermal behaviour of okra (*Abelmoschus esculentus*) fibres. Compos Sci Technol 71:246–254

De Rosa IM, Kenny JM, Puglia D, Santulli C, Sarasini F (2010) Morphological, thermal and mechanical characterization of okra (*Abelmoschus esculentus*) fibres as potential reinforcement in polymer composites. Compos Sci Technol 70(1):116–122

Fortunati E, Puglia D, Monti M, Santulli C, Maniruzzaman M, Kenny JM (2013a) Cellulose nanocrystals extracted from okra fibers in PVA nanocomposites. J Appl Polym Sci 128:3220–3230

Fortunati E, Puglia D, Monti M, Santulli C, Maniruzzaman M, Foresti ML, Vazquez A, Kenny JM (2013b) Okra (*Abelmoschus esculentus*) fibre based PLA composites: mechanical behaviour and biodegradation. Journal Polym Environ 21:726–737

Franklin WM (1982) Okra, potential multiple-purpose crop for the temperate zones and tropics. Econ Bot 36(3):340–345

Gassan J, Bledzki AK (1998) Alkali treatment of jute fibers : relationship between structure and mechanical properties. J Appl Polym Sci 623–629

Hamideh H, Mohini S, Lucia HM (2014) Modification and characterization of hemp and sisal fibers. J Nat Fibers 11(2):144–168

Islam MS, Pickering KL (2014) An empirical equation for predicting mechanical property of chemically treated natural fibre using a statistically designed experiment. Fibers Polym 15(2):355–363

Khan G, Alam M (2013) Surface chemical treatments of jute fiber for high value composite uses. Res Rev J Mater Sci 1(2):39–44

Kumar DS, Tony DE, Praveen KA, Kumar KA, Srinivasa RDB, Nadendla R (2013) A review on: *Abelmoschus esculentus* (okra). Int Res J Pharm Appl Sci 3(34). www.irjpas.com

Li X, Lope A, Tabil G, Panigrahi S (2007) Chemical treatments of natural fiber for use in natural fiber-reinforced composites: a review. J Polym Environ 15:25–33

Moniruzzaman M, Maniruzzaman M, Gafur MA, Santulli C (2009) Lady's finger fibres for possible use as a reinforcement in composite materials. J Biobased Mater Bioenergy 3:1–5

Mwaikambo LY, Ansell MP (2002) Chemical modification of hemp, sisal, jute, and kapok fibers by alkalization. J Appl Polym Sci 84(12):2222–2234

Onyedum O, Aduloju SC, Sheidu SO, Metu CS, Owolabi OB (2016) Comparative mechanical analysis of okra fiber and banana fiber composite used in manufacturing automotive car bumpers. Am J Eng Technol Soc 2(6):193–199

Srinivasababu N (2015) An overview of okra fibre reinforced polymer composites. In: IOP Conference Series: Materials Science and Engineering, vol 83

Srinivasababu N, Murali Mohan Rao K, Suresh Kumar J (2009) Tensile properties characterization of okra woven fiber reinforced polyester composites. Int J Eng 3(4):403–412

Van Soest PJ, Wine RH (1968) Determination of lignin and cellulose in acid-detergent fiber with permanganate. J Assoc Official Anal Chem 51:780–785

Yılmaz ND, Konak S, Yılmaz K, Kartal AA, Kayahan E (2016) Characterization, modification and use of biomass: okra fibers. Bioinspired Biomim Nanobiomaterials 5(3):85–95

Printed in the United States
By Bookmasters